SIMPLE BENDING

Other titles in this series

*

Force Systems and Equilibrium
J. G. A. CROLL

Stress and Strain
T. R. GRAVES SMITH

Virtual Work and Energy Concepts
J. RHODES

Torsion
A. C. WALKER

The Buckling of Struts
A. C. WALKER

Modular Textbooks in Engineering

Editor: A. C. WALKER, Ph.D.
Department of Civil and Municipal Engineering,
University College, London

Simple Bending

J. R. TYNE
D.L.C., B.Sc.(Eng.), M.I.C.E., M.I.Struct.E.
Department of Civil Engineering
Polytechnic of Central London

1974

CHATTO & WINDUS
LONDON

Published by
Chatto & Windus Ltd
42 William IV Street
London WC2N 4DF
*
Clarke, Irwin & Co. Ltd
Toronto

ISBN 0 7011 1953 5

Printed in Great Britain by
T. & A. Constable Ltd
Hopetoun Street
Edinburgh EH7 4NF

CONTENTS

1

INTRODUCTION

Bending is a deformation which consists of lateral displacements of the longitudinal axis of a member, usually as a result of the action of lateral forces or couples. The purpose of this book is to present the theories used by engineers to relate these applied forces and couples with the internal stresses and the deformations they produce.

The type of member whose bending is considered is that in which the lateral dimensions are relatively small compared with the length, and for which the bending displacements are also small. In practice such members occur frequently in building and engineering structures, for example the timber floor joists of a domestic house or the aluminium alloy spars of an aeroplane wing. In nature the stems of plants and the trunks and branches of trees exhibit the same structural behaviour.

1.1 Development of bending theories

In early days engineers decided on the dimensions of timber beams and stone lintels on the basis of their own and others' experience, and a considerable degree of refinement of design was achieved (presumably by trial and error), if one is to judge by contemporary records and by many of the historic buildings and other constructions which are still standing today. It appears that the first attempt to provide a rational basis for beam design came from Galileo in the early 17th century, who conducted tests on stone cantilevers. He was concerned with the *strength* of the cantilevers rather than their deformations (which would have been very small anyway) and he concluded that failure occurred with the whole of the cross section in tension, rotating about the lower edge. After the discovery of the fundamental principle of elasticity by Hooke in 1660, Mariotte, in 1680, came to the conclusion that bending results in the extension of some of the longitudinal 'fibres' of a beam and the compression of others, and that there existed a layer at which neither occurred, at the mid height of the section. In 1705, James Bernouilli derived an equation for the curve of a bent rod, from the extension and compression of the longitudinal fibres, and in 1776 Coulomb completed the analysis of bending in terms of longitudinal stresses in the form used by engineers today. This analysis attributes the deformation entirely to the longitudinal stresses, the fundamental proposition being that plane cross sections are not distorted and remain plane during bending.

The objections to the Coulomb theory, on the basis of the mathematical theory of elasticity, are that it takes no account of shear stresses and that distortion of the cross section must obviously occur. In the late 1830s St Venant produced a rigorous analysis of bending in which these factors were taken into account, and which gave direct stresses in accord with the Coulomb theory. All the earlier theories were for special cases, such as a member with terminal couples, or a cantilever with an end load, and the first application of these theories to the general problem of the bending of a beam subjected to lateral loads was made by Navier in 1833. Navier also successfully tackled the problem of statically indeterminate beams. Although the St Venant analysis gave the distribution of shear stress, it was difficult to apply to all except a few simple cases, and so an approximate shear stress analysis by Jourawski in 1854, backed up by tests on timber beams, is the one which has been used by engineers ever since.

Many of the early theories of bending were supported by little or no direct experimental work, and tended to fall within the province of mathematics or philosophy. However, much experimental knowledge was gained from the tests conducted by British engineers in the mid-1800's to enable them to design the structures required by the expansion of industry, particularly railway structures. These tests were more concerned with the properties of engineering materials and the behaviour of structural elements than with the development of new theories. Hodgkinson, in 1830, published the results of tests on beams made from cast iron, from which he concluded that the material has different elastic moduli for tension and compression. He also studied the failure in bending of cast iron I section beams, in order to find the optimum proportions of the tension and compression flanges.

In the present century the design of structures in various materials, such as structural streel and reinforced concrete, has been mainly based on the early bending theories, and there has been a continual improvement both in material properties and in the methods of structural analysis. In recent years attention has been directed to the behaviour of members in bending as failure is approached, when the linear elastic basis of the classical bending theories no longer applies. In particular, the post-yield bending of steel and reinforced concrete have been examined experimentally and theoretically and the conclusion drawn that the assumption of plane sections remaining plane is still valid. Hence the analysis differs only in the stress/strain characteristics of the materials.

Since 1845, when Fairbairn investigated the proposals for a new railway bridge over the Conway, the problem of the buckling of girders

constructed from thin plates has been of concern to engineers. Design practice has tended to be based on the linear elastic bending theory, with restrictions on working stresses in regions prone to buckling and the provision of stiffening members to inhibit this mode of failure. The recent collapses of several box-girder bridges has stimulated further research into this problem.

1.2 Scope of the book

The first topic is independent of any theory of bending, being based on purely statical considerations. The distribution and variation of the stress resultants of bending are considered for various types of member and load systems.

For the determination of stresses in Chapter 3 the classical theories are developed and applied to a variety of linearly elastic problems, including composite sections. Non-linear elastic and plastic bending are also studied, together with the general case of linearly elastic bending of non-symmetrical sections.

The last chapter deals with displacements, on the basis of the linear elastic theory, and the work extends to include some statically indeterminate problems.

It should be clearly understood that the bending theories presented here are approximations which are found to be sufficiently accurate for most engineering design purposes, for members whose length considerably exceeds their lateral dimensions. For a more precise analysis reference should be made to the mathematical theory of elasticity.*

* A. E. H. Love, *The Mathematical Theory of Elasticity* (Dover Publications), 4th edition, 1944; S. Timoshenko and J. N. Goodier, *The Theory of Elasticity* (McGraw-Hill).

2

STRESS RESULTANTS

2.1 Bending in one plane

Taking the simplest case to start with, we deal here with members and loading systems which lie in a single plane, and whose geometry is not significantly affected by deformations.

The member ABC is loaded with forces and couples such as W and m, and internal stresses are thereby generated; typical stresses acting on the part of the member to the right of a normal section at B are shown in (b). In (c) all such stresses acting over the whole cross section have been replaced by statically equivalent forces P and Q together with a couple M, the *stress resultants*.

In (d) is shown the part of the member BC which is in equilibrium under the action of these stress resultants at B together with the external force W and couple m.

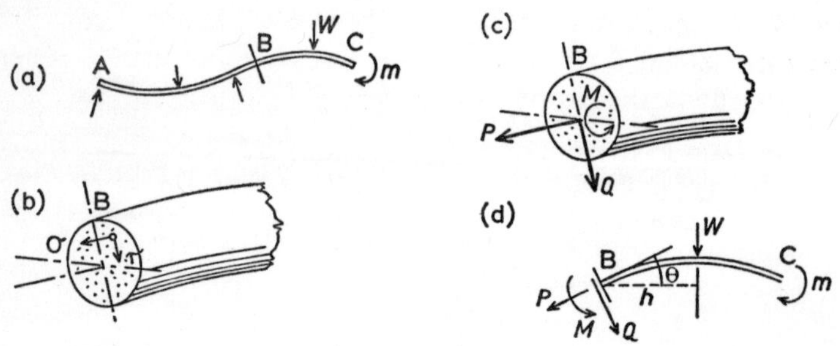

Fig. 2.1

Applying the conditions for statical equilibrium we find:

$$P = -W \sin\theta \quad \text{(resolving axially at B)},$$

$$Q = -W \cos\theta \quad \text{(resolving normally at B)},$$

$$M = Wh + m \quad \text{(taking moments about B)}.$$

Clearly, the same numerical values for P, Q, M would have resulted had we investigated the equilibrium of the part of the member AB (and in practice one would choose whichever part gave the least computation!), but the signs would have been different. In other words, the stress resultants occur in equal and opposite pairs, and may be evaluated as follows:

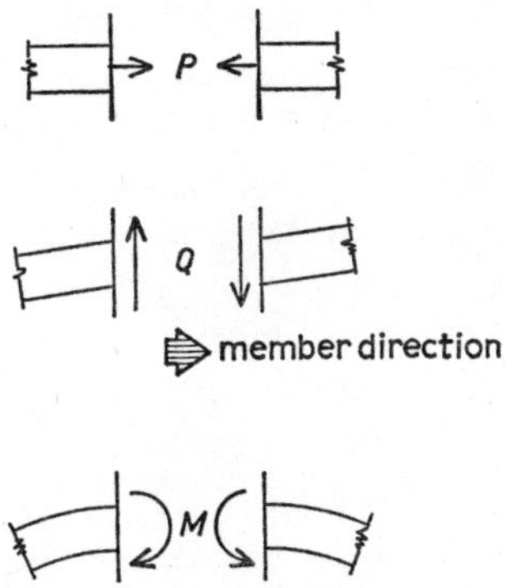

Fig. 2.2

Axial force is equal and opposite to the axial component of the resultant of the forces acting to one side of a section.

Shear force is equal and opposite to the normal component of the resultant of the forces acting to one side of a section.

Bending moment is equal and opposite to the couple resultant of the forces and couples acting to one side of a section.

The sign conventions used for the stress resultants do not follow universally recognised hard and fast rules and are a matter of preference and convenience, varying from author to author and indeed from application to application. The system used in this text, by which the resultants shown in Fig. 2.2 are all taken as *positive*, has the following significant merits:

1. The signs for axial force and shear force are consistent with those for the corresponding stresses used in stress analysis.
2. The sign convention for bending moment is consistent with the widely accepted graphical representation of bending moments on the *tension* side of a member, and is also consistent with the right-hand screw rule for moments with respect to the X-axis of the cross section of a member.
3. The sign conventions for shear force and bending moment are mutually compatible in that there is no need for a minus sign in the differential relation between them, nor, indeed, in this text, in any of the further differential relations involving slopes, displacements, etc.

It should be noted that shear force, being a skew symmetrical function, will appear to have a different sign according to which side of the member is being viewed. To avoid ambiguity, therefore, it is necessary to fix which way the member is viewed, and this may most conveniently be achieved by specifying the member direction, left to right. Axial force and bending moment are symmetrical functions and so this problem does not arise.

The three stress resultants described above refer to a general loading system for which deformations other than bending will also occur. The axial force is clearly associated only with longitudinal extension or shortening, and it is the *shear force and bending moment which are the stress resultants of bending*.

It should be made clear at this stage that the foregoing definitions of the stress resultants are adequate only if the lateral dimensions of the member are small compared with the length, since the precise points of action of the shear force and axial force within the section have not been defined. For members whose lateral dimensions are not small it is necessary to define the former, for instance, such that the axial force acting alone at this point would not give rise to any bending. Now bending is a phenomenon of deformation, and so it is not sufficient to ensure that *stresses* are uniform or symmetrical over the section, since non-uniformity or non-symmetry of material properties might still result in lateral displacements of the longitudinal axis of the member. When Chapter 3, concerned with stresses, has been studied it will be appreciated that for linearly elastic materials this point is the centroid of the section, but that for other material properties other points are appropriate A further important requirement is that this point must lie in the plane of the applied loading at all positions throughout the length of a member for the resulting displacements to be within this plane. However, for most practical cases, where the lateral dimensions are reasonably small, it is sufficiently accurate to work to the centre line of the member. For further discussion of these aspects see § 2.2.

EXAMPLE 2.1

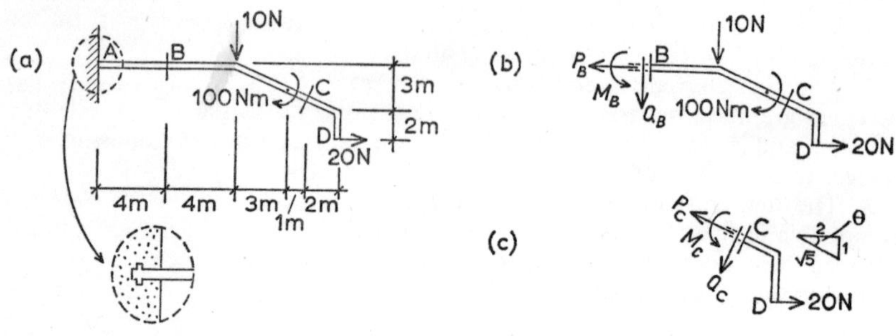

Fig. 2.3

The member ABCD is built in to a rigid body at A so that all displacements and rotation at this point are prevented. The member is described as being *fully fixed* or *encastré* at A and, being completely free at the other end, is known as a *cantilever*.

The problem is to evaluate the stress resultants at B and C.

To find the stress resultants at B, consider the equilibrium of the part of the member BD shown separately in (b),

resolving axially at B, the axial force $P_B = 20$ N,

resolving normally at B, the shear force $Q_B = -10$ N,
taking moments about B, the bending moment

$$M_B = 10\times4+100-20\times5 = 40 \text{ Nm}.$$

To find the stress resultants at C, consider the equilibrium of the part of the member CD shown separately in (c),
resolving axially at C, the axial force $P_C = 20 \cos\theta = 8\sqrt{5}$ N,
resolving normally at C, the shear force $Q_C = 20 \sin\theta = 4\sqrt{5}$ N,
taking moments about C, the bending moment $M_C = -20\times3 = -60$ Nm.

EXAMPLE 2.2

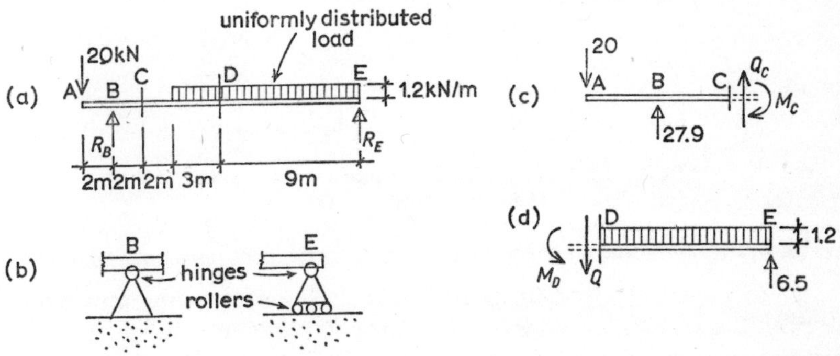

Fig. 2.4

The beam ABCDE is supported at B and E in such a way that horizontal reactions are not generated. To achieve this and yet at the same time 'locate' the member in the horizontal direction, one of the supports should be capable of providing a horizontal restraint, the member being free to move in this direction at the other. A suitable arrangement is shown diagramatically in (b), but of course such ideal 'simple support' conditions are rarely found in practice where beams rest on non-frictionless bearings or are fixed to other members. However, the assumption of 'simple support' can be made with sufficient accuracy in most practical cases where the horizontal restraint does not generate significant secondary stresses.

(Note the convention, used throughout this book, by which applied forces are represented by open arrows and reactions by closed ones.)

Distributed loading occurs more frequently in practice than any other, where beams support slabs or walls and are subjected to gravity loads or wind pressures, etc.

The problem is to evaluate the stress resultants at C and D.

It is first necessary to calculate the reactions at B and E by considering the statical equilibrium of the member as a whole;
taking moments about B,

$$R_E\times16 = 1.2\times12\times10-20\times2, \text{ therefore } R_E = 6.5 \text{ kN},$$

resolving vertically,

$$R_B+R_E = 1.2\times12+20, \qquad \text{therefore } R_B = 27.9 \text{ kN}.$$

Since the beam is subjected to normal forces only, the stress resultants at any point will consist of shear force and bending moment only.

Stress resultants at C

The simplest arithmetic will result from considering the part of the member to the left of C, shown separately in (c);

$$Q_C = 20 - 27.9 = -7.9 \text{ kN},$$

$$M_C = 20 \times 4 - 27.9 \times 2 = 24.2 \text{ kNm}.$$

Stress resultants at D

For D it is easier to consider the part to the right, shown separately in (d);

$$Q_D = 6.5 - 1.2 \times 9 = -4.3 \text{ kN},$$

$$M_D = 1.2 \times 9 \times 4.5 - 6.5 \times 9 = -9.9 \text{ kNm}.$$

(check that the same values would have resulted from considering the part of the member to the left. Watch the signs of M and Q!)

(a) *Bending moment and shear force diagrams for rectilinear members*

In this section we investigate the shear force and bending moment at all positions along a member for one particular loading arrangement. For rectilinear members this information is conveniently presented graphically by means of diagrams in which the values of shear force or bending moment are set off perpendicularly to the axis of the member. These diagrams are of obvious value to an engineer who is designing the member to carry safely the bending moment and shear force at every point. With experience, and making use of standard cases and results obtained later on in this section, it is possible to produce these diagrams very rapidly for the systems of point and uniformly distributed loading most frequently occurring in practice.

Rectilinear members which are subjected primarily to loads normal to their longitudinal axis are referred to as beams, or as joists or girders in various practical contexts, and bending is the most significant phenomenon in the behaviour of such members. However, axial forces and deformations may in practice occur which are either of a secondary nature, such as those due to expansion or shrinkage, or are of prime importance such as longitudinal prestressing, and in both these cases the axial effects must be taken into account in design. We shall begin by considering various beam problems from the point of view of the bending effects only since in the majority of cases it is possible to deal with these effects separately from any others. When we come to non-rectilinear members in § 2.1(d), the distribution of axial force will be considered as well as that of the bending moment and shear force.

EXAMPLE 2.3

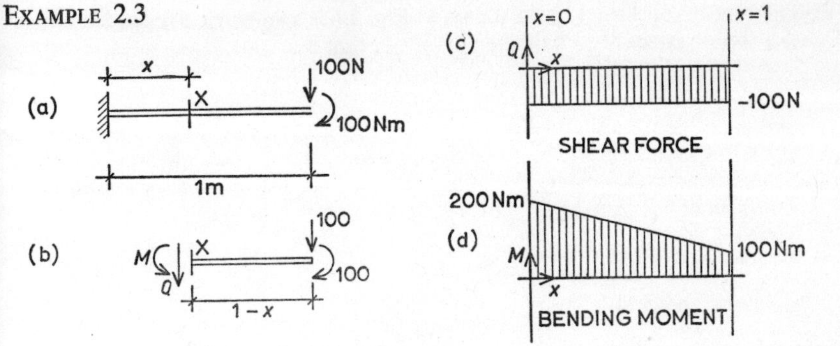

Fig. 2.5

For the simple cantilever shown in Fig. 2.5 the bending moment and shear force on a section at a general position X distance x from the left-hand end will be continuous functions of the variable x, valid for all positions between $x = 0$ and $x = 1$.

Consider the equilibrium of the part of the member shown separately in (b), to the right of the section at the general position X,

resolving vertically,

$$Q = -100 \text{ N},$$

moments about X,

$$M = 100+100(1-x) = (200-100x) \text{ Nm}.$$

These expressions have been plotted for all values of x in (c), the SHEAR FORCE DIAGRAM and (d), the BENDING MOMENT DIAGRAM.

EXAMPLE 2.4

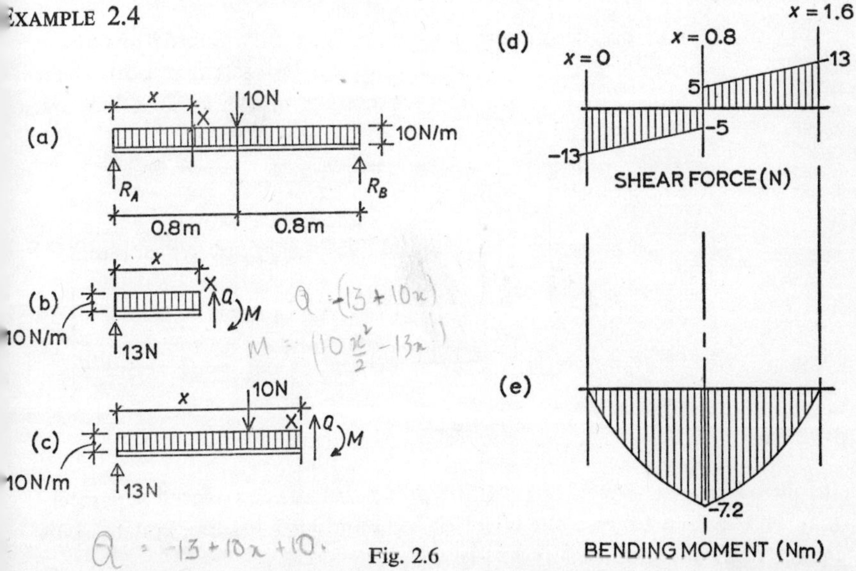

Q = −13 + 10x + 10

Fig. 2.6

M =

The simply supported beam shown in Fig. 2.6 is symmetrically loaded and so each reaction is equal to half the total load, that is

$$R_A = R_B = \tfrac{1}{2}(10+10\times 1.6) = 13\text{N}.$$

In this problem we shall have to consider two general positions, one to the left of the point load and one to the right, and so obtain two expressions for shear force and bending moment. For values of x between 0 and 0.8 we consider the equilibrium of the part of the beam shown separately in (b),

resolving vertically,

$$Q = 10x-13,$$

moments about X,

$$M = 10x\times\frac{x}{2}-13x.$$

For values of x between 0.8 and 1.6 the part of the beam shown separately in (c) is considered,

resolving vertically,

$$Q = 10x+10-13 = 10x-3,$$

moments about X,

$$M = 10x\times\frac{x}{2}+10(x-0.8)-13x.$$

These expressions have been plotted in (d) and (e), and the discontinuity caused by the point load can be clearly seen. Note that the symmetrical loading produces a symmetrical bending moment diagram but a skew-symmetrical shear force diagram.

(b) *Superimposition of shear force and bending moment diagrams*

For the type of problem being considered here, in which deformations do not significantly affect the geometry, it can be seen that both shear force and bending moment are linear functions of the applied loading.

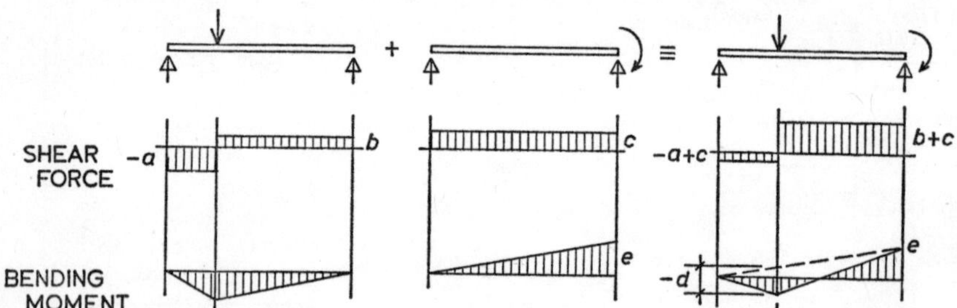

Fig. 2.7

It follows that the effects of several loading systems considered separately may be summed to give the effect of the combined loading system. This is illustrated diagramatically in Fig. 2.7.

This property is particularly useful in more complicated problems where the shear force and bending moment diagrams may be constructed by combining the results for various components of the loading system, perhaps making use of well-known standard cases. A number of simple but

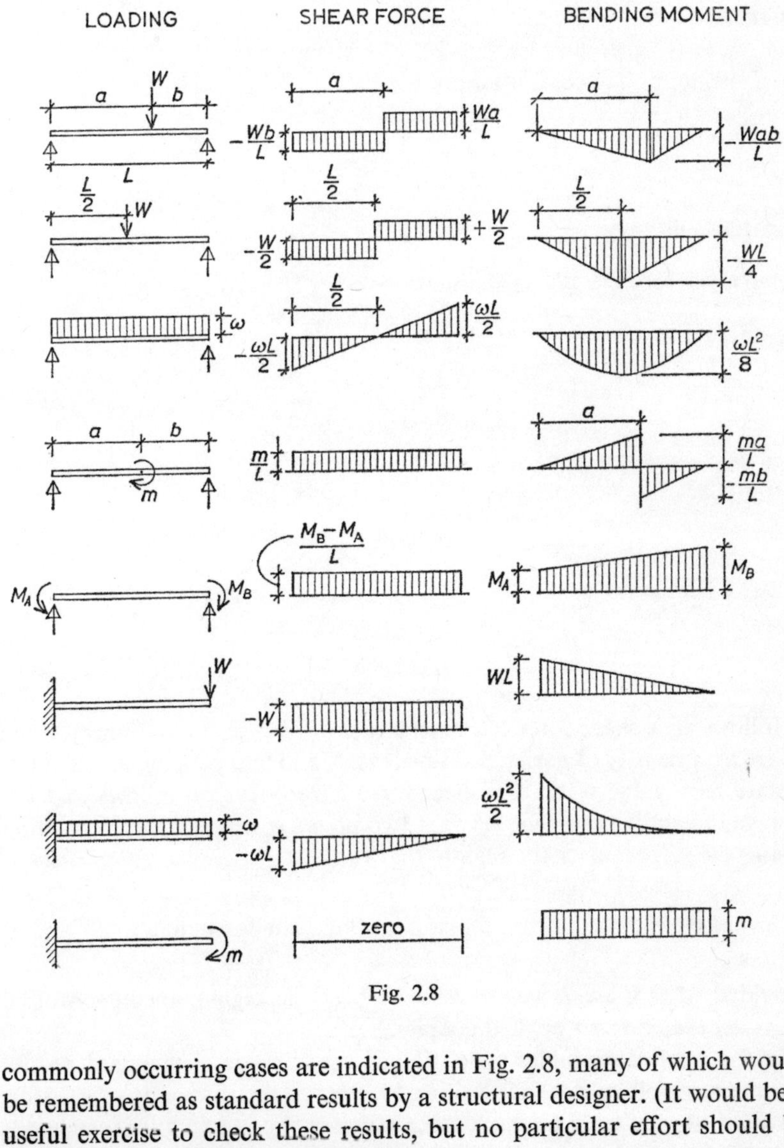

Fig. 2.8

commonly occurring cases are indicated in Fig. 2.8, many of which would be remembered as standard results by a structural designer. (It would be a useful exercise to check these results, but no particular effort should be made to commit them to memory.)

(c) *The relation between shear force, bending moment and the intensity of normal loading*

Fig. 2.9 shows an elemental length of member dx, at a point where the intensity of normal loading is ω. Considering the equilibrium of this element,

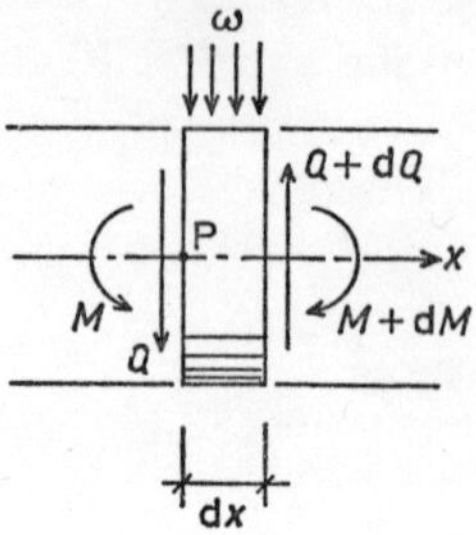

Fig. 2.9

resolving normal to the axis,

$$Q+\omega \mathrm{d}x = Q+\mathrm{d}Q,$$

therefore

$$\frac{\mathrm{d}Q}{\mathrm{d}x} = \omega, \tag{2.1}$$

and taking moments about P

$$(M+\mathrm{d}M)+(\omega \mathrm{d}x)\frac{\mathrm{d}x}{2} - (Q+\mathrm{d}Q)\mathrm{d}x = M.$$

In the limit as $dx \to 0$,

$$\frac{\mathrm{d}M}{\mathrm{d}x} = Q. \tag{2.2}$$

It follows that shear force has a zero rate of change, or stationary value where the intensity of normal loading is zero, and that bending moment has a stationary value where the shear force is zero. It also follows that the change in bending moment between two points is equal to the summation of the shear force over the region, that is

$$M_B - M_A = \int_A^B Q.\mathrm{d}x, \text{ the area of the shear force diagram,} \tag{2.3}$$

provided M is a *single valued function* over the region. In other words, provided there are no applied couples.

In the worked examples which follow use is made of standard results and the principle of superimposition together with the results of § 2.1(c). It would be profitable to check these examples by working independently from first principles.

EXAMPLE 2.5

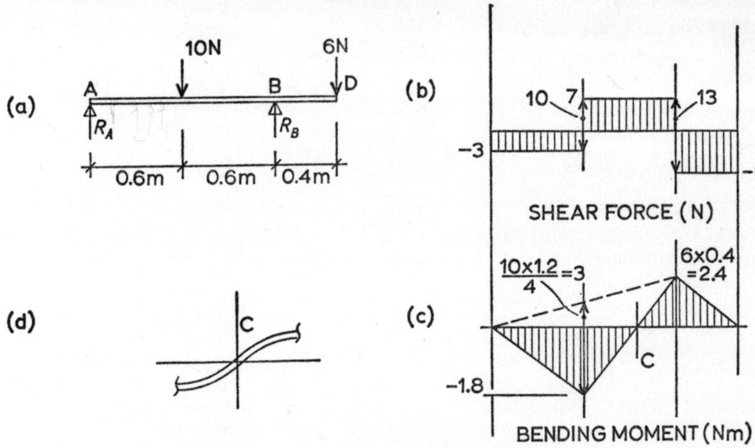

Fig. 2.10

Consideration of the equilibrium of the beam as a whole yields the following values for the reactions,

$$R_B = 13, \quad R_A = 3.$$

It will be noted that the shear force is constant in regions free of load and that the step discontinuities correspond to the magnitudes of the point loads and reactions. The bending moment diagram has been constructed by combining the standard result for a simple beam with a central point load with the bending moments produced by the load at D.

The point C at which the bending moment changes sign is known as a *point of contraflexure*, on account of the corresponding reversal of curvature of the beam centre line as indicated in (d), see Chapter 4.

EXAMPLE 2.6

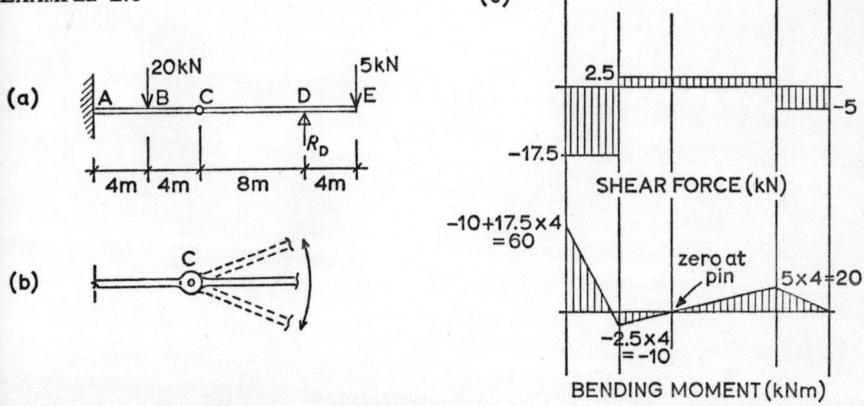

Fig. 2.11

The beam ABCDE in Fig. 2.11 has a pin joint at C which is illustrated in (b). There can obviously be no bending moment at such a joint, but it is capable of taking shear forces. In practice, articulated beam arrangements are frequently found in bridges where the joint will often also make provision for horizontal movement to allow for expansion.

The reaction R_D is found by writing the condition that the bending moment at C is zero,

$$R_D \times 8 = 5 \times 12, \text{ therefore } R_D = 7.5 \text{ kN}.$$

It is not necessary to evaluate the reaction and fixing moment at A since the shear force and bending moment can everywhere be calculated in terms of the forces to the right of a section.

Example 2.7

In Fig. 2.12 the reactions R_B and R_C are found by considering the equilibrium of the beam as a whole,

moments about B,

$$R_C = \frac{1 \times 8 \times 4 + 2 \times 10 - 4 \times 3}{8} = 5 \text{ kN},$$

resolving vertically,

$$R_B = 4 + 1 \times 8 + 2 - R_C = 9 \text{ kN}.$$

In this example the standard result for a beam with a uniformly distributed load has been combined with the bending moments due to the loads at A and D.

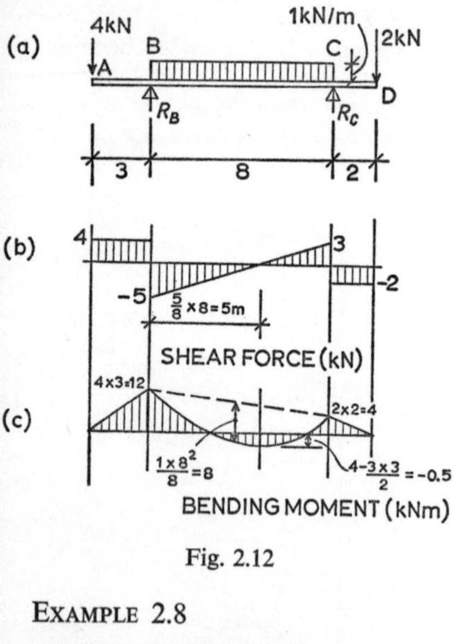

Fig. 2.12

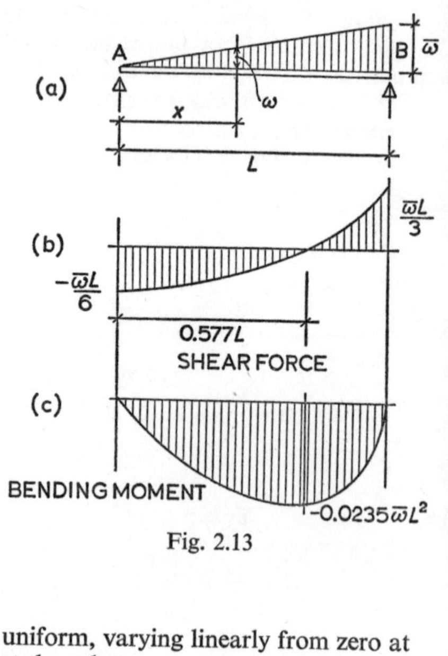

Fig. 2.13

Example 2.8

In Fig. 2.13 the distributed load is not uniform, varying linearly from zero at one support to the maximum value of $\bar{\omega}$ at the other.

Since the intensity of normal loading is a continuous function of distance along the beam we may make use of equations (2.1) and (2.2) to obtain, by integration, functions for the shear force and bending moment.

At any point, distance x from the left-hand support, the intensity of loading is given by $\omega = \dfrac{\varpi x}{L}$.

From equation (2.1) we have $\dfrac{dQ}{dx} = \omega$, which may be integrated to give

$$Q = \frac{\varpi x^2}{2L} + C_1,$$

where C_1 is a constant of integration.

From equation (2.2) we have $\dfrac{dM}{dx} = Q$, which may be integrated to give

$$M = \frac{\varpi x^3}{6L} + C_1 x + C_2,$$

where C_2 is a second constant of integration.

The values of the constants C_1 and C_2 may now be found by reference to the boundary conditions of the problem,

at A, M is zero, therefore $\dfrac{\varpi}{6L}(0)^3 + C_1(0) + C_2 = 0$ and $C_2 = 0$,

at B, M is also zero, giving $\dfrac{\varpi L^3}{6L} + C_1 L = 0$, whence $C_1 = -\dfrac{\varpi L}{6}$.

The expressions for shear force and bending moment then become,

$$Q = \frac{\varpi x^2}{2L} - \frac{\varpi L}{6}, \quad M = \frac{\varpi x^3}{6L} - \frac{\varpi L x}{6},$$

and these have been plotted in Fig. 2.13 (b) and (c).

A stationary value of the bending moment will occur when the shear force is zero, that is when $\dfrac{\varpi x^2}{2L} - \dfrac{\varpi L}{6} = 0$,

$$\text{giving } x = \frac{L}{\sqrt{3}} = 0.577L.$$

At this position the bending moment is $\dfrac{\varpi L^2}{18\sqrt{3}} - \dfrac{\varpi L^2}{18} = -0.0235\varpi L^2$.

(d) *Bending moment and shear force diagrams for non-rectilinear members*

In many applications the members subjected to bending are anything but rectilinear, or are built up of rectilinear components many of which are not horizontal. It is apparent that the simple sign convention which served for horizontal beams will need to be extended to enable the stress resultants to be described uniquely in these cases.

It will be remembered (p. 11) that since shear force is a skew-symmetrical pair of resultants it was necessary to associate a notional 'member direction' with the positive sign. If in the case of a non-rectilinear member we assign a direction from one end to the other, or to each part if the member is composed of several components, it will be possible to use the sign convention for shear force and bending moment without ambiguity.

It follows that the sign attached to a particular shear force or bending moment will be dependent on the notional direction chosen for the member, but it should be noted that when bending moments are plotted normal to the axis following this convention *they are always found to be on the tension side*. This feature provides a very convenient way of remembering and interpreting the convention, and is very helpful to a designer who appreciates a clear graphical indication of the nature of the bending stresses.

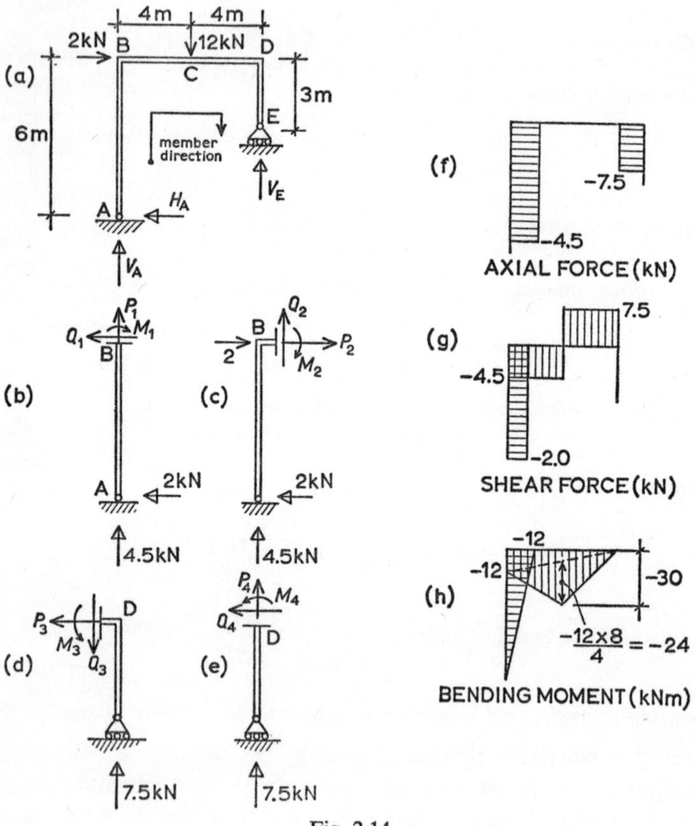

Fig. 2.14

The analysis of multi-storey frameworks and of most types of portal frames and arches falls outside the scope of this book, but we shall consider two statically determinate examples to illustrate the principles. In these examples the axial force distribution will also be determined since from a practical point of view this is often of considerable significance.

EXAMPLE 2.9

The portal frame ABCDE in Fig. 2.14 is supported by a hinge at A and by a hinge on rollers at E. There are thus only three reactions and these may be determined by considering the equilibrium of the frame as a whole,

moments about A,

$$V_E = \frac{2\times 6+12\times 4}{8} = 7.5 \text{ kN},$$

resolving vertically,

$$V_A = 12-7.5 = 4.5 \text{ kN},$$

resolving horizontally,

$$H_A = 2 \text{ kN}.$$

It will only be necessary to evaluate the stress resultants at the ends of each of the rectilinear parts of this frame, by considering the equilibrium of the portions shown separately in (b), (c), (d) and (e). The member direction will be taken as ABCDE.

The stress resultants are found as follows:

(b) $P_1 = -4.5$ kN

$Q_1 = -2$ kN

$M_1 = -2\times 6 = -12$ kNm

(c) $P_2 = 0$

$Q_2 = -4.5$ kN

$M_2 = -2\times 6 = -12$ kNm

(d) $P_3 = 0$

$Q_3 = 7.5$ kN

$M_3 = 0$

(e) $P_4 = -7.5$ kN

$Q_4 = 0$

$M_4 = 0$

The diagrams (f), (g) and (h) have been plotted using these values and the standard results for a member with a central point load (BCD).

EXAMPLE 2.10

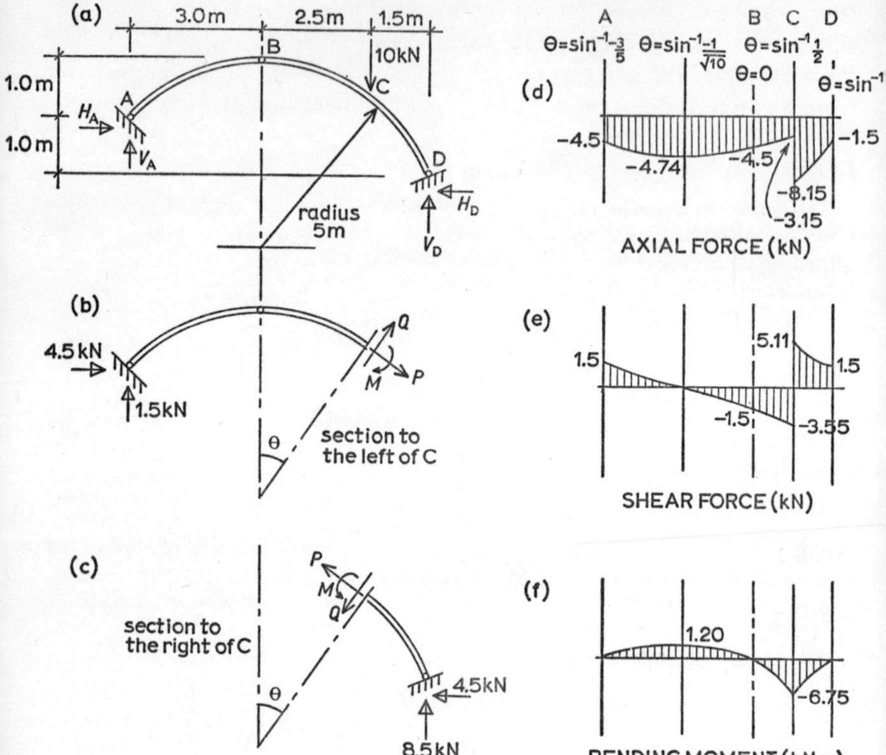

Fig. 2.15

The three-pinned arch rib ABCD has four reactions, but the pin joint at the crown provides an additional equilibrium condition over and above the three relating to the equilibrium of the arch as a whole. The reactions are found as follows:

taking moments about D for the whole arch,

$$V_A \times 7 + H_A \times 1 = 10 \times 1.5;$$

the bending moment at B is zero, therefore

$$V_A \times 3 - H_A \times 1 = 0.$$

From these two equations,

$$H_A = 4.5 \text{ and } V_A = 1.5 \text{ kN}.$$

Resolving horizontally for the whole arch,

$$H_D = H_A = 4.5 \text{ kN},$$

resolving vertically,

$$V_D = 10 - 1.5 = 8.5 \text{ kN}.$$

It would not be practicable in this case to plot the stress resultants by setting them off perpendicular to the axis of the member at each point. The resultants will be evaluated in terms of the variable θ and will be plotted on a horizontal base. We shall have to consider two general positions, and the parts of the member to one side of sections at each of these positions are shown separately in (b) and (c). Considering the equilibrium of these parts we obtain the following results,

(b) Axial force $P = -4.5 \cos \theta + 1.5 \sin \theta$.
Shear force $Q = -4.5 \sin \theta - 1.5 \cos \theta$.
Bending moment $M = 4.5(5 \cos \theta - 4) - 1.5(3 + 5 \sin \theta)$.

(c) Axial force $P = -4.5 \cos \theta - 8.5 \sin \theta$.
Shear force $Q = 8.5 \cos \theta - 4.5 \sin \theta$.
Bending moment $M = 4.5(5 \cos \theta - 3) - 8.5(4 - 5 \sin \theta)$.

These results are plotted in (d), (e) and (f).

PROBLEMS

(answers are given at the end of the chapter)

Plot shear force and bending moment diagrams for the following members, locating and determining stationary values of bending moment where these occur.

PROBLEM 2.1

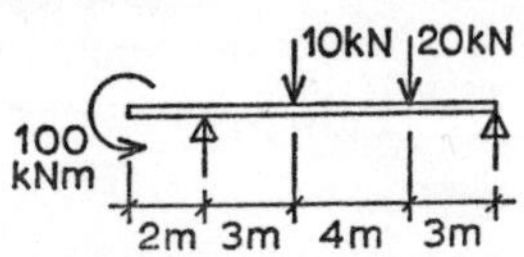

PROBLEM 2.2

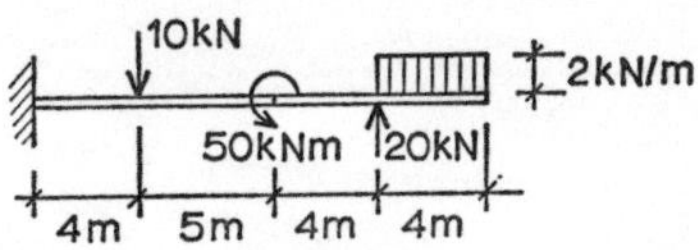

PROBLEM 2.3

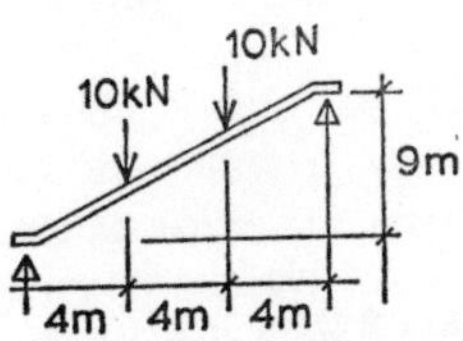

(plot on a horizontal base)

PROBLEM 2.4

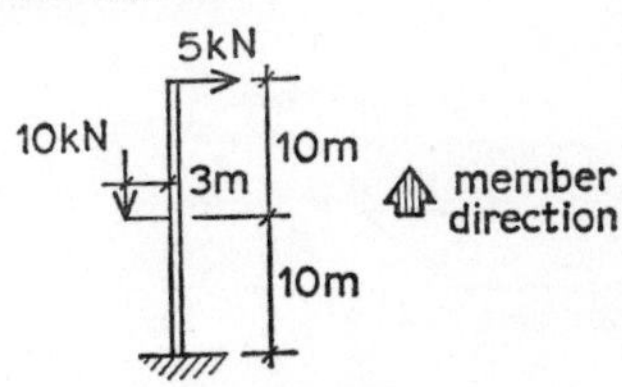

PROBLEM 2.5. *A prestressed beam*

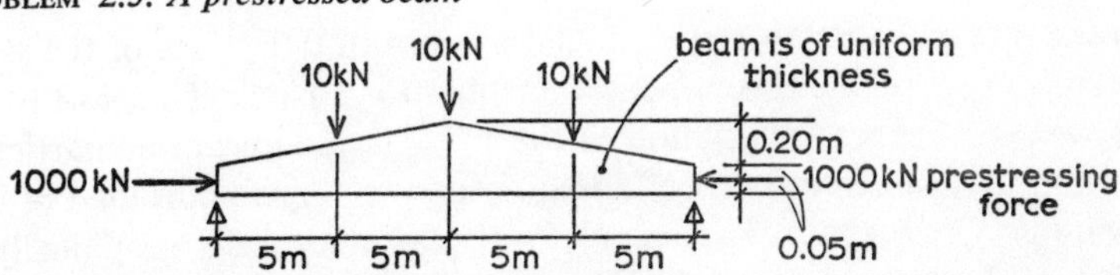

(Remember the significance of the point of action of the axial force; in this case use the centroid. It is useful to consider the effects of the vertical loads and of the prestressing force separately.)

PROBLEM 2.6

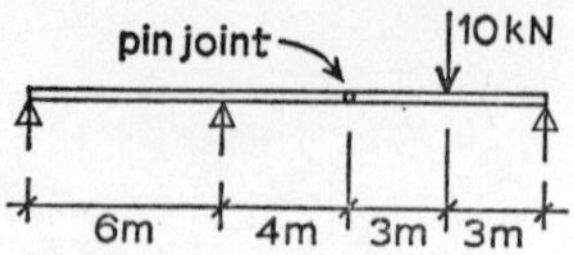

PROBLEM 2.7

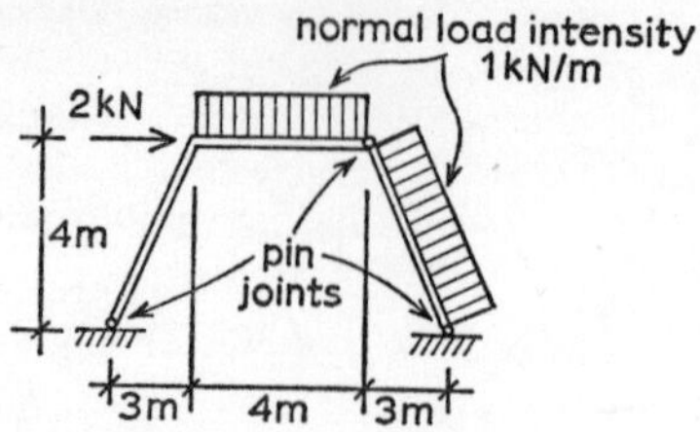

PROBLEM 2.8

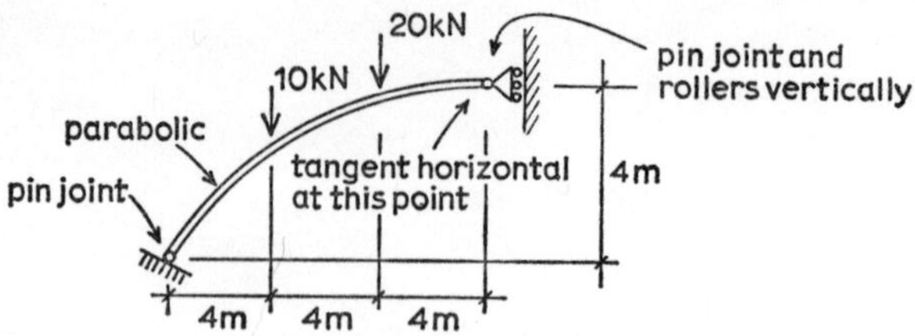

(Plot all three stress resultants on a horizontal base.)

PROBLEM 2.9

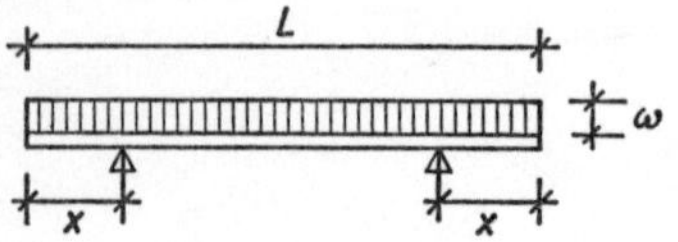

What value of x will give the least numerical value of maximum bending moment and what will this bending moment be?

PROBLEM 2.10

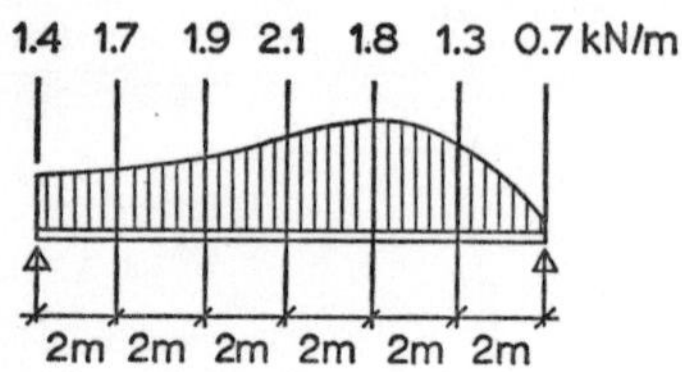

The beam is subjected to a non-uniform distributed load of the intensities shown. (Start from

$$\frac{dQ}{dx} = \omega$$

and use numerical integration, or replace the loading by an equivalent system of point loads.)

(e) *Influence lines for bending moment and shear force*
(The following is an introduction to the basic ideas only)

In the previous section we were concerned with the values of the stress resultants at all points throughout a member for a particular set of loads. Now we investigate the variation of the stress resultants at one particular position for various positions of a load, usually a single point load of unit magnitude. Once again it is useful to display the results graphically in diagrams in which the ordinate represents the stress resultant, but this time it is the stress resultant *at a single position*. The base is the axis of load position. These diagrams are known as *influence lines*, showing the influence of load position on the value of the stress resultant at a particular point, and are of obvious value when designing structures subjected to moving loads such as bridges or crane girders.

EXAMPLE 2.11

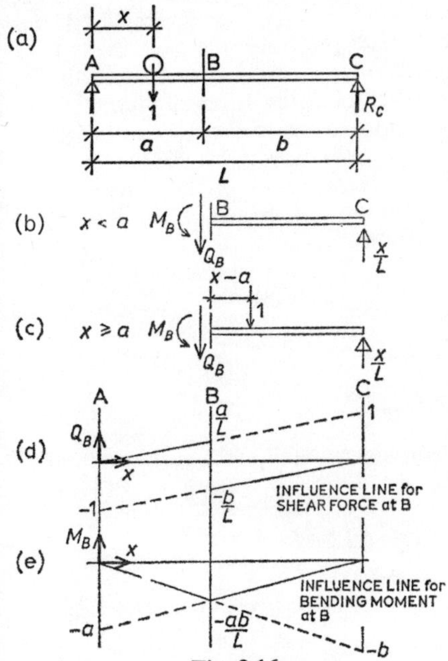

Fig. 2.16

The problem is to investigate the shear force and bending moment at B as x is varied. It is clear that different expressions will be obtained according as x is less or greater than a. For all values of x, however, there will be a single expression for the reaction R_C,

$$R_c = 1 \times \frac{x}{L}.$$

The part of the beam to the right of a section at B is shown in (b) for x less than a and in (c) for x greater than a. Evaluating the stress resultants in each case we find,

$$\text{for } x<a, \quad Q_B = \frac{x}{L}, \quad M_B = -\frac{xb}{L},$$

$$\text{for } x \geqq a, \quad Q_B = -1+\frac{x}{L},$$

$$M_B = 1\times(x-a) - \frac{xb}{L}.$$

These results are plotted in (d) and (e).

EXAMPLE 2.12

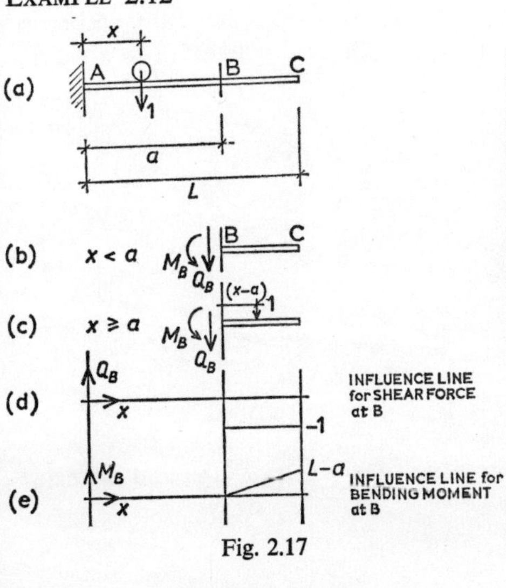

Fig. 2.17

The problem is to obtain influence lines for shear force and bending moment at B. As in the previous example we obtain different expressions for the stress resultants according as x is less or greater than a, and diagrams (b) and (c) show the part of the cantilever to the right of a section at B for these two cases.

For $x<a$, $Q_B = 0$ and $M_B = 0$.

For $x \geqq a$, $Q_B = -q$

and $M_B = 1\times(x-a)$.

These results are plotted in (d) and (e).

EXAMPLE 2.13

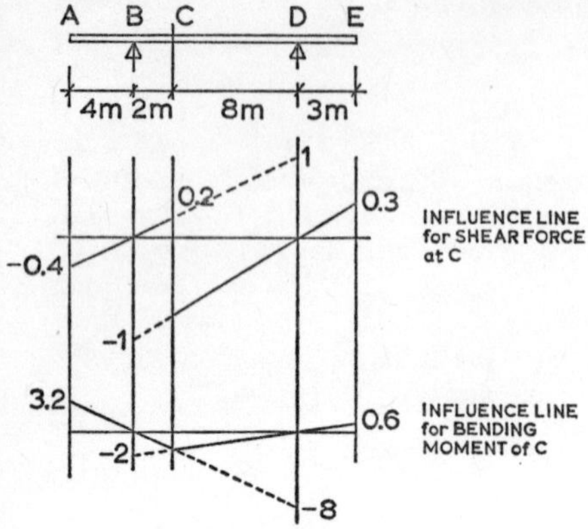

Fig. 2.18

The influence lines for shear force and bending moment at C will be similar to those in Example 2.11 for load positions between B and D. However, there will be no discontinuity in the stress resultant expressions as the load moves to the left of B, or to the right of D. Hence the influence lines in these regions are simply extensions of the inner portions.

PROBLEMS

(*answers are given at the end of the chapter*)

Plot influence lines for shear force and bending moment at the positions marked X in the following beams.

PROBLEM 2.11

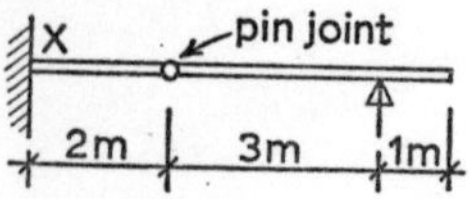

PROBLEM 2.12

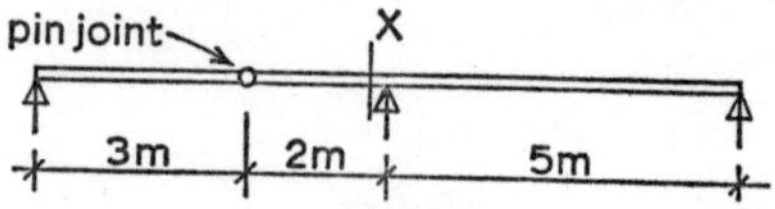

2.2 The general case

In this section we consider the general problem of a flexural member in which the stress resultants of bending do not everywhere fall in a single

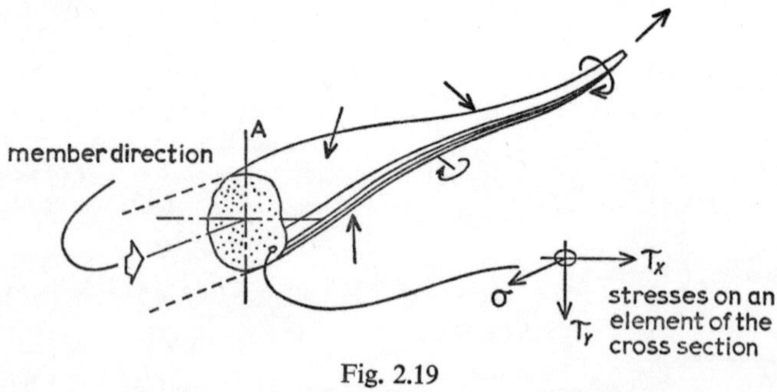

Fig. 2.19

plane, in other words, when a member of arbitrary shape and material properties is loaded arbitrarily.

Fig. 2.19 shows part of a member, subjected to various forces and couples, lying to one side of a normal cross section at A. The section will be subject to direct and shearing stresses as shown in the smaller diagram. As in § 2.1 the equilibrium of this part of the member is unaffected by replacing these stresses by a statically equivalent set of forces and couples, the 'stress resultants'. These stress resultants may act at any convenient point within the section and are equal and opposite to those which apply to the other part of the member. The stress resultants acting on both parts are shown in Fig. 2.20, six in number, consisting of the 'bending' resultants *M* and *Q* with respect to both the axes *X* and *Y*, together with the axial force, *P*, and the *torque T*.

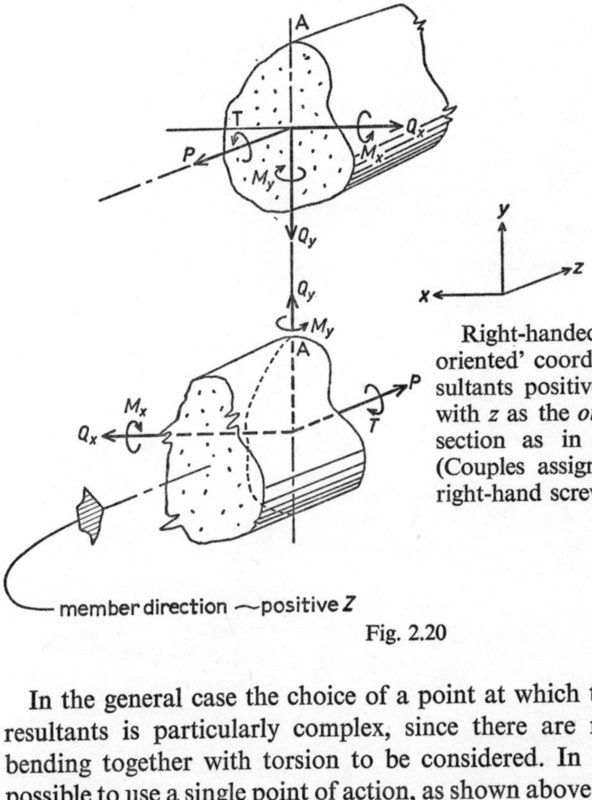

Right-handed set of 'member oriented' coordinate axes. Stress resultants positive in these directions with *z* as the *outward* normal to the section as in the lower diagram. (Couples assigned according to the right-hand screw rule.)

Fig. 2.20

In the general case the choice of a point at which to express the stress resultants is particularly complex, since there are now two planes of bending together with torsion to be considered. In fact, it may not be possible to use a single point of action, as shown above, at all. For instance, in the problem of the bending and twisting of a linearly elastic, non-symmetrical section it is convenient to have the shear forces passing through the 'shear centre'* and the axial force through the centroid. The magnitudes of the forces are not affected by the choice of position, but those of

* *Mechanics of Elastic Structures*, J. T. Oden (McGraw Hill) 1967

the couples are. However, as stated in § 2.1, for sections whose lateral dimensions are small compared with the length, the distinction is of no great importance, and for the common case of symmetrical sections of homogeneous linearly elastic material both the bending and the twisting criteria lead to the same point (the centroid) anyway.

These criteria are such as to make it possible to isolate the different effects for consideration separately, as in § 2.1 where it was stated that the axial force is positioned such that it would give rise, on its own, to no bending (on the basis of the *actual material properties which exist for the combined problem*). The shear forces should be positioned so that, acting alone, they would give rise to no torsion.

EXAMPLE 2.14

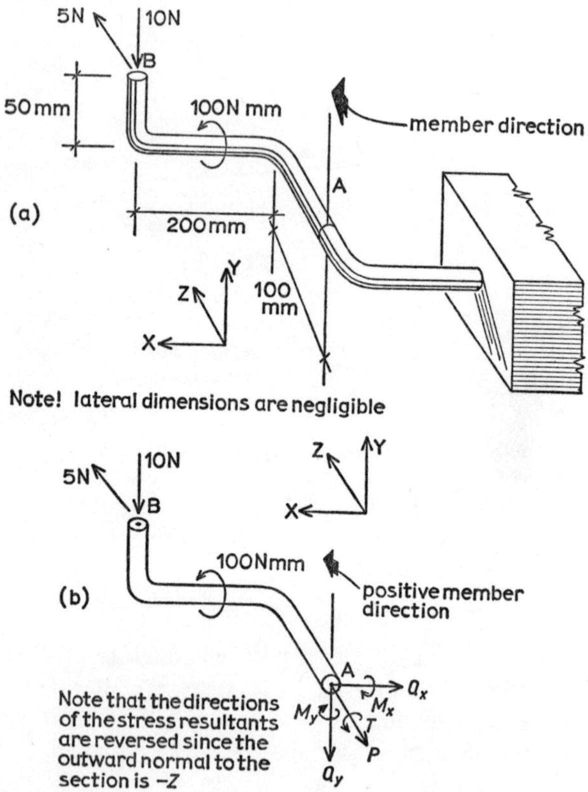

Fig. 2.21

In order to determine the stress resultants at A, consider the part of the member beyond A, shown separately in Fig. 2.21 (b);

For equilibrium of AB we have,

$$P = 5 \text{ N}$$
$$T = -10 \times 200 = -2000 \text{ Nmm}$$
$$Q_x = 0$$
$$Q_y = -10 \text{ N}$$
$$M_x = 100 + 5 \times 50 + 10 \times 100 = 1350 \text{ Nmm}$$
$$M_y = -5 \times 200 = -1000 \text{ Nmm.}$$

Study of the previous example should make it clear that force components which affect the shear force Q_y appear in the expression for M_x, but not M_y. Similarly Q_x is related to M_y but not M_x. (It must not, however, be assumed that the resultant of Q_x and Q_y would coincide with the direction of the resultant of M_x and M_y.)

This means that the results established in § 2.1 and all the subsequent work in that section relating to bending in a single plane may all be applied to each of the axes of bending X and Y individually, and that an arbitrary loading system on a rectilinear member may be resolved into the two planes with each considered separately for purposes of distribution diagrams, influence lines, etc. For non-rectilinear members there is in general no single 'plane of bending' applicable to every part of the member, and so the work of § 2.1 may not be applicable. Certain non-planar members do possess a convenient basis for the description of the stress resultant distribution, such as those having a simply definable shape like the coil spring. Another class of problem which presents no difficulty is that in which the non-planar structure is built up from individually rectilinear members (such as in example 2.14), in which case the results of § 2.1 may be applied, once the coordinate axes of each member have been defined. For the general arbitrary member, however, no generalisation is possible, and a practical approach might be to approximate by substituting an 'equivalent' member made up of short rectilinear pieces.

ANSWERS TO PROBLEMS 2.1-2.12

2.1

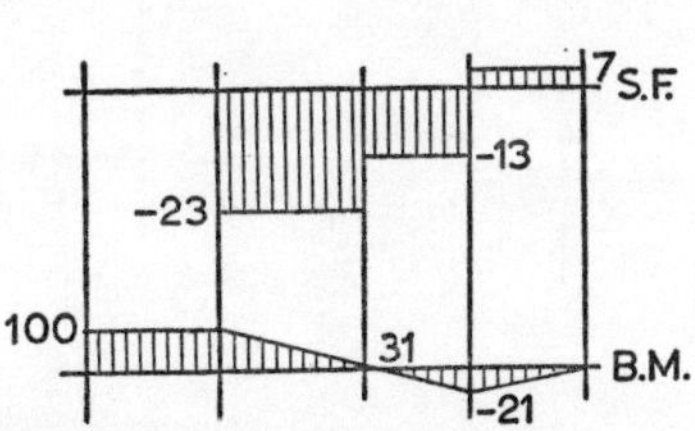

2.2

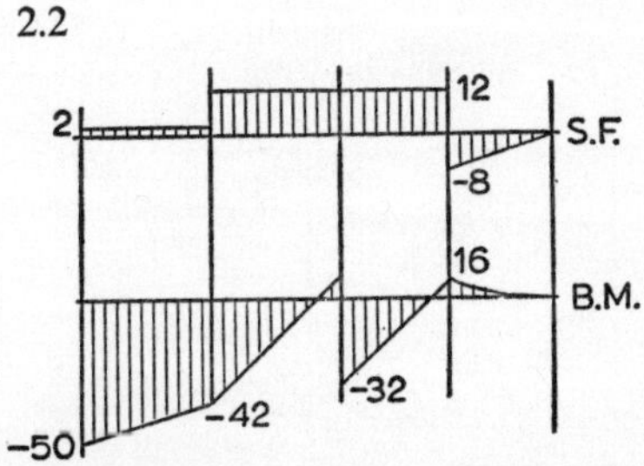

2.3

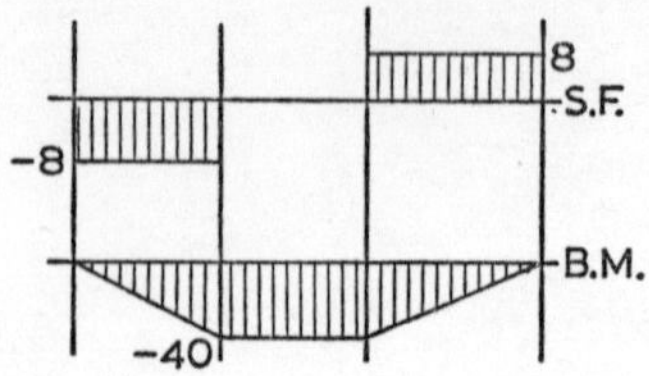

2.4

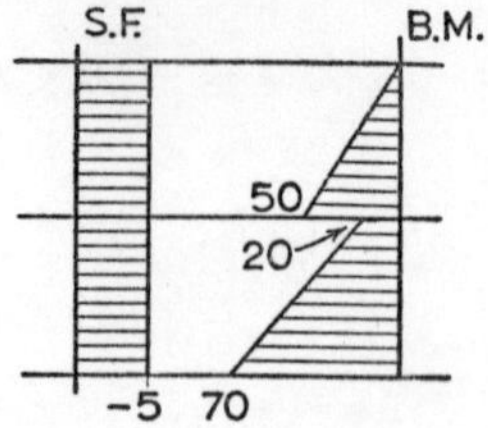

2.5

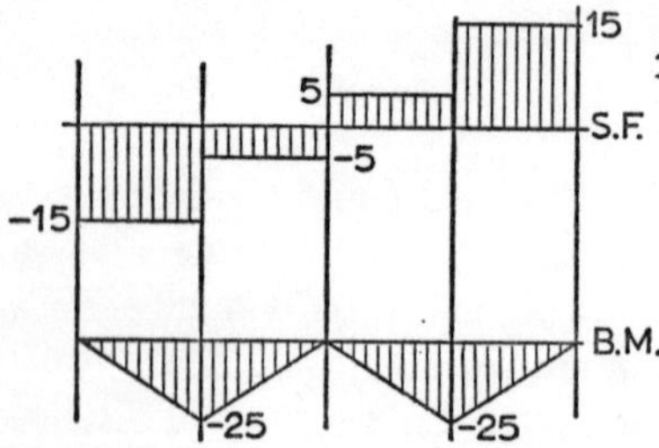

2.6

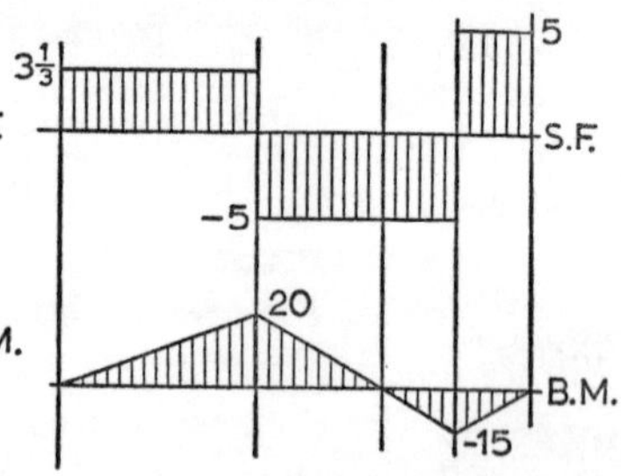

2.7

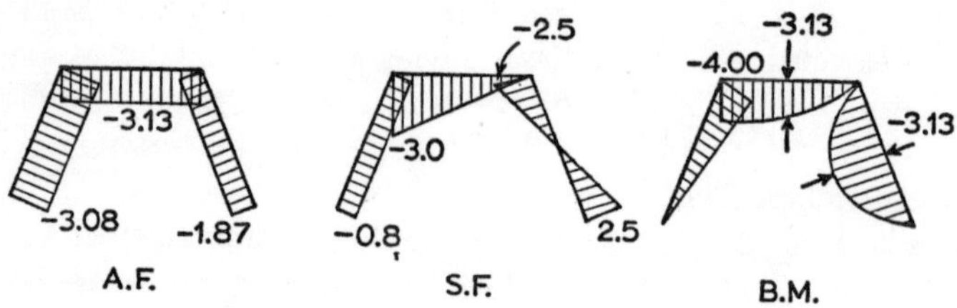

2.8

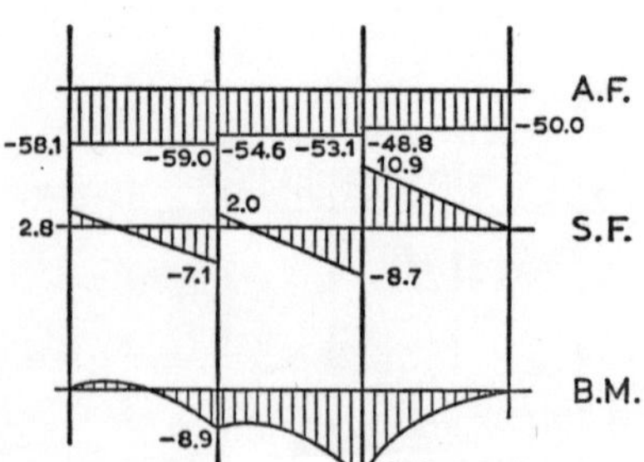

2.9

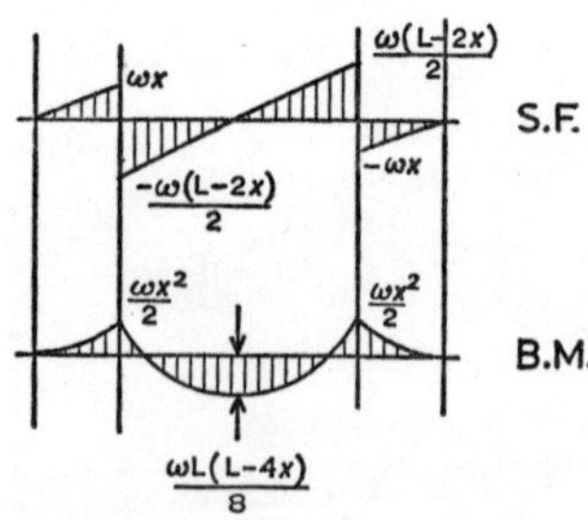

$x = 0.207L$

Max BM $= \pm 0.0215\omega L$

2.10

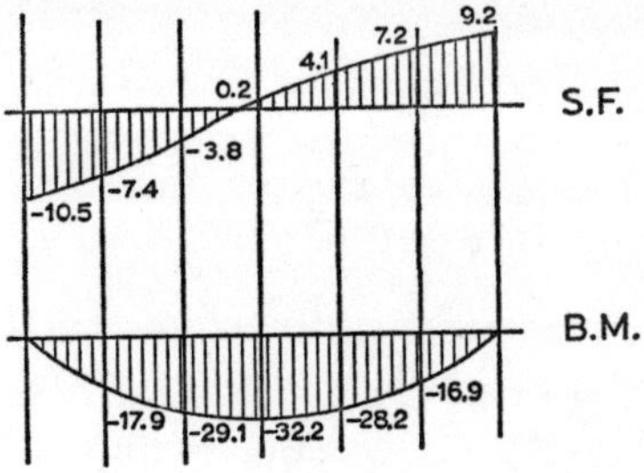
9.2
7.2
4.1
0.2
-3.8
-7.4
-10.5
S.F.
B.M.
-16.9
-17.9
-29.1
-32.2
-28.2

2.11

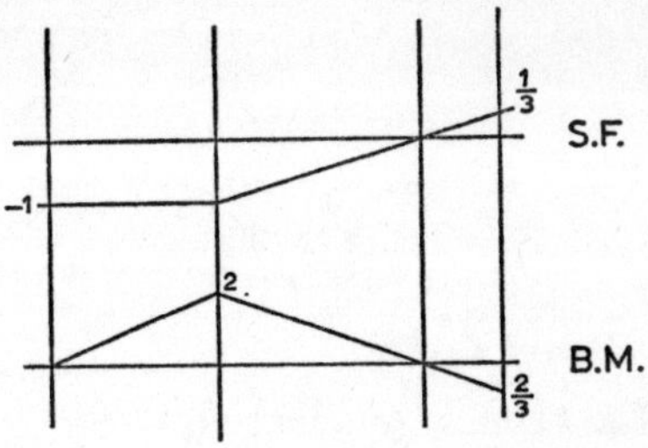
1/3
S.F.
-1
2
B.M.
2/3

2.12

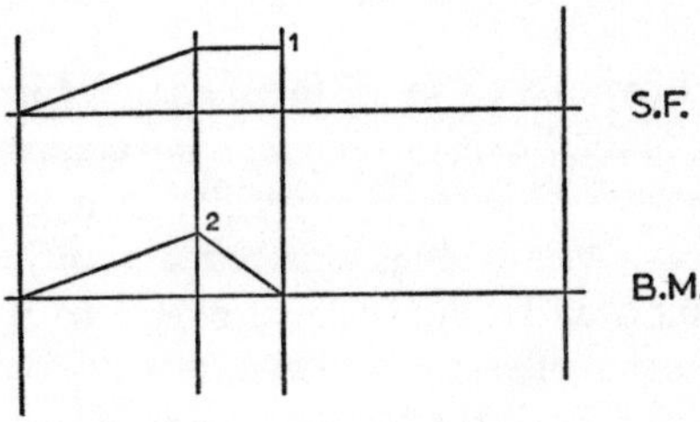
1
S.F.
2
B.M.

3

INTERNAL STRESSES

3.1 Direct stresses

The bending moment on a section is the couple resultant of the normal (direct) stresses. In the following we consider the magnitudes and distribution of these stresses for various shapes and orientation of the section with respect to the applied bending moment, and for various material properties.

The fundamental hypothesis in all these analyses is that plane sections normal to the longitudinal axis of the member remain plane during bending, and this gives results which compare well with experiment for members whose lateral dimensions are small compared with their length, and for areas sufficiently removed from the local effects of things such as point loads or abrupt changes of section.

There will in practice be distortion and warping of the cross section, but these effects are found to be of secondary importance for most practical purposes.

(a) *Symmetrical sections of linearly elastic material*

When a member of symmetrical cross section is subjected to bending moments acting in the plane of symmetry, the resulting displacements are also in (or parallel to) this plane. In other words, plane cross sections of the member remain perpendicular to the plane of symmetry. In this analysis the member is assumed to be initially straight, or at least of a radius of curvature which is large compared with the lateral dimensions.

Fig. 3.1 shows a thin slice of the member between two adjacent plane sections, which have rotated relative to each other as the member has bent, under the action of the applied moments, to a (local) radius R. (There is also a displacement associated with the shear forces, consideration of which is outside the scope of this work, but for members whose lateral dimensions are small compared with their length, these displacements are relatively insignificant. In a conventional steel joist section, for instance, the displacement resulting from shear forces would be at most one or two per cent of the flexural displacement.) For an infinitesimally thin slice the line of intersection of the plane sections may be taken as the axis of curvature of the member. There being no resultant normal forces acting on the sections it follows that part of the slice is compressed and part extended either side of a zero strain level, and from the symmetry of the problem it may be concluded that this 'layer' appears on the cross section as a straight line, known as the *Neutral Axis*, perpendicular to the axis of symmetry.

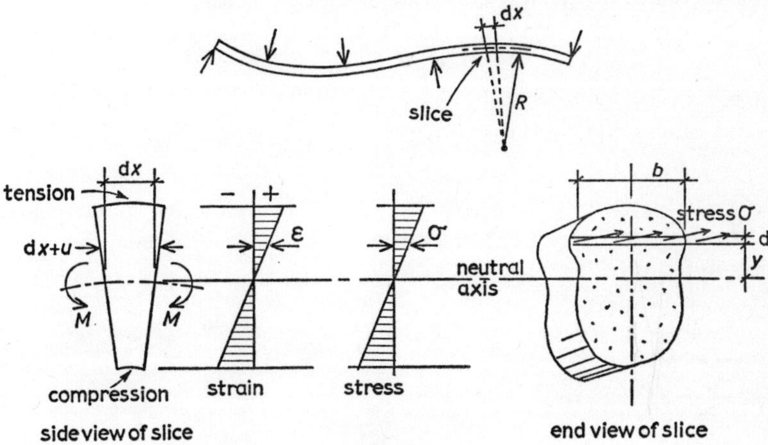

Fig. 3.1

The increase in length of the layer y above the neutral axis is given by,

$$u = \frac{(R+y)}{R}\,\mathrm{d}x - \mathrm{d}x = \frac{y\mathrm{d}x}{R}.$$

Hence the strain is $\varepsilon = \dfrac{u}{\mathrm{d}x} = \dfrac{y}{R}.$

For a linear elastic material, with Young's modulus denoted by E, the stress σ at the layer y is

$$\sigma = E\varepsilon = \frac{Ey}{R}. \tag{3.1}$$

The total longitudinal force on the layer, thickness $\mathrm{d}y$, is given by

$$\sigma b\mathrm{d}y = \frac{E}{R}\,yb\mathrm{d}y.$$

There is no resultant longitudinal force on the section, so that

$$\int_{\text{section}} \frac{E}{R}\,yb\mathrm{d}y = 0.$$

Since $\dfrac{E}{R}$ is constant for this integration we have

$$\int_{\text{section}} yb\mathrm{d}y = 0.$$

This equation expresses the condition that the first moment of area of the section about the neutral axis is zero, that is, *the neutral axis passes through the centroid of the section.* The resultant of the longitudinal forces on the layers is a couple equal to the applied bending moment M. Taking moments about the neutral axis for convenience gives

$$\int_{\text{section}} \left(\frac{E}{R}\,yb\mathrm{d}y\right) y = M,$$

that is,

$$\frac{E}{R}\int_{\text{section}} y^2 b\mathrm{d}y = M.$$

The integral is the *second moment of area I* (often erroneously referred to as the 'moment of inertia') of the section about the neutral axis. (See notes on second moments of area below.)

Thus we can write

$$\frac{EI}{R} = M, \tag{3.2}$$

and combining (3.1) and (3.2) gives

$$\sigma = \frac{My}{I}. \tag{3.3}$$

Obviously maximum positive and negative values of σ correspond to maximum positive and negative values of y. The quantity $\dfrac{I}{y_{\max}}$ is called the *section modulus Z* giving

$$\sigma_{\max} = \frac{M}{Z}. \tag{3.4}$$

NOTES ON SECOND MOMENTS OF AREA

Standard cases

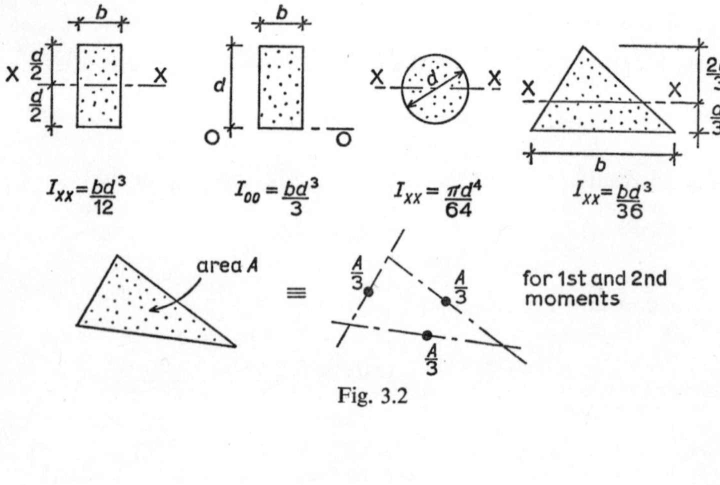

Fig. 3.2

Parallel axes theorem

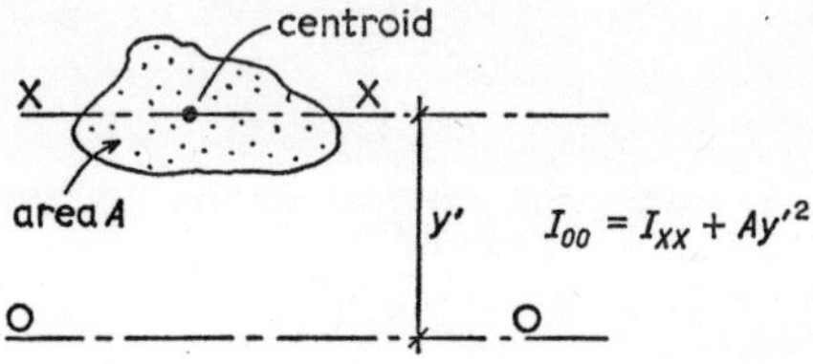

Fig. 3.3

EXAMPLE 3.1

Find the position of the neutral axis and determine the section moduli for the section shown. (All dimensions in millimetres.)

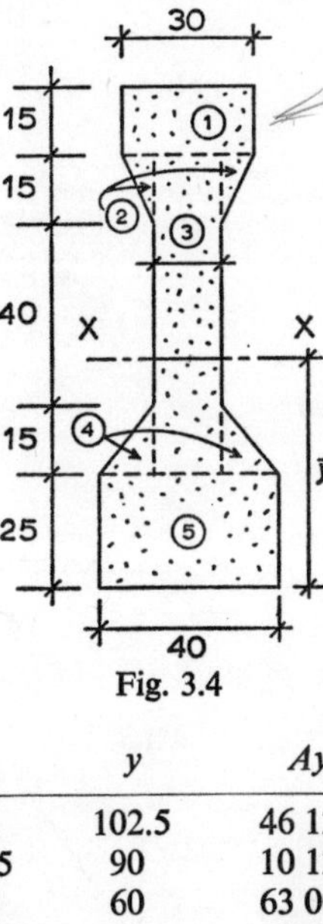

Fig. 3.4

In the table below:

A is the area of a numbered part
y is the ordinate of the centroid of a part above the base.
y' is the ordinate of the centroid of a part above the neutral axis X-X.
I_G is the second moment of area of a part about an axis through its own centroid.

Part	A	y	Ay	y'	I_G	Ay'^2
1	450	102.5	46 125	53.5	8 450	1 290 000
2	112.5	90	10 125	41.0	1 405	189 000
3	1 050	60	63 000	11.0	428 750	121 605
4	187.5	30	5 625	−19.0	2 344	67 600
5	1 000	12.5	12 500	−36.5	52 083	1 333 000

$$\Sigma A = 2\,800, \quad \Sigma Ay = 137\,375, \quad \Sigma I_G = 493\,032, \quad \Sigma Ay'^2 = 3\,001\,205$$

$$\therefore \quad \bar{y} = \frac{137\,375}{2\,800} = 49.0 \text{ mm}$$

and $$I_{xx} = \Sigma I_G + \Sigma Ay'^2 = 3\,494\,000 \text{ mm}^4.$$

Hence the section moduli are:

Top $$Z_t = \frac{I}{y_t} = 57\,300 \text{ mm}^3$$

Bottom $$Z_b = \frac{I}{y_b} = 71\ 300\ \text{mm}^3.$$

EXAMPLE 3.2

A joist, Fig. 3.5, is rectangular in cross section, 200 mm × 75 mm, and is subjected to a bending moment of 1 kNm. What is the maximum stress produced?

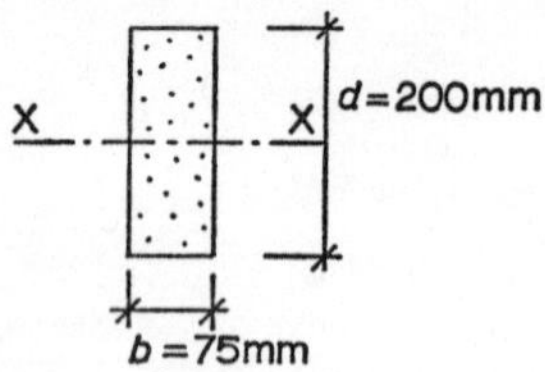

Fig. 3.5

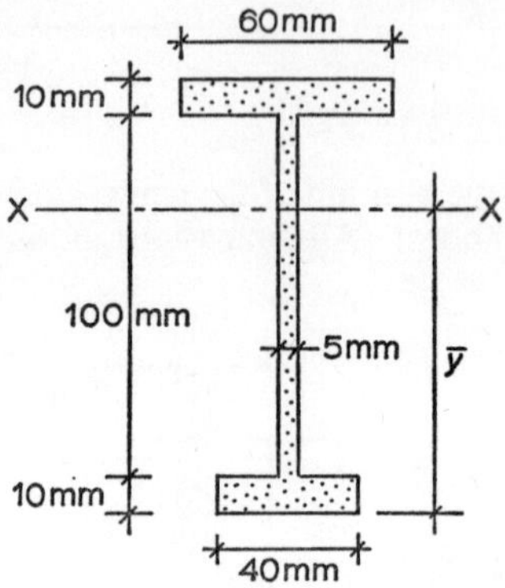

Fig. 3.6

$$I_{xx} = \frac{bd^3}{12}, \quad z = \frac{I}{\frac{1}{2}d} = \frac{bd^2}{6}. \quad \text{(Top and bottom)}$$

Therefore $$\sigma_{max} = \pm\frac{1\ 000 \times 1\ 000 \times 6}{75 \times 200^2} = \pm 2\ \text{N/mm}^2.$$

EXAMPLE 3.3

The bending stress at the top of the section shown in Fig. 3.6 is +100 N/mm². What is it at the bottom?

$$\bar{y} = \frac{(600 \times 115) + (500 \times 60) + (400 \times 5)}{(600 + 500 + 400)} = 67.3\ \text{mm}.$$

Since stresses are proportional to distance from the neutral axis, the stress at the bottom is given by,

$$\sigma_b = (-)100 \times \frac{67.3}{(120 - 67.3)} = -128\ \text{N/mm}^2.$$

PROBLEMS

PROBLEM 3.1

The cross section of a flexural member is an equilateral triangle with its base horizontal. If the maximum permissible tensile stress is one-quarter of the maximum compressive stress, what is the ratio of the maximum permissible positive and negative bending moments? (1:2)

PROBLEM 3.2

Find the maximum bending stress set up when a simple joist of 2 500 mm span is traversed by two point loads of 10 kN, 500 mm apart. The section modulus of the joist is 150×10^3 mm³. (66.7 N/mm²)

PROBLEM 3.3

What bending moment is required to bend an initially straight strip, 50 mm × 1 mm thick, to a radius of 1 m, if the modulus of elasticity is 10^5 N/mm²? (417 N/mm)

PROBLEM 3.4

A circular section tube whose internal and external diameters are 10 and 11 mm respectively, is simply supported on a span of 200 mm. If the specific weight of the material is 100 kN/m³, what is the maximum bending stress set up? (0.199 N/mm²)

PROBLEM 3.5

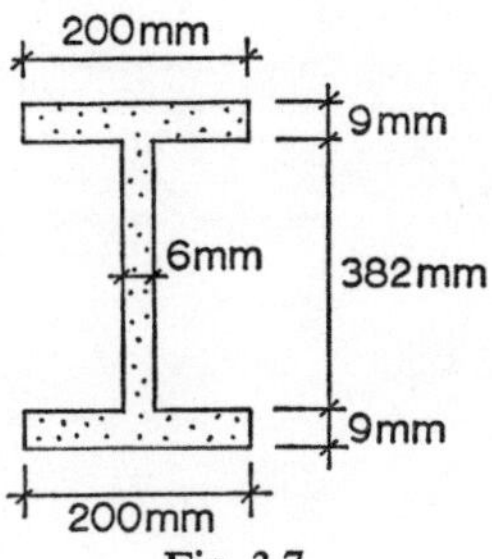

Fig. 3.7

What proportion of the bending moment on the section is taken by the flanges of the beam shown? (83%)

PROBLEM 3.6

Compare the section moduli of the following sections with that of a solid square having the same area. (1.414:1, 0.844:1, 1.44:1, 0.707:1)

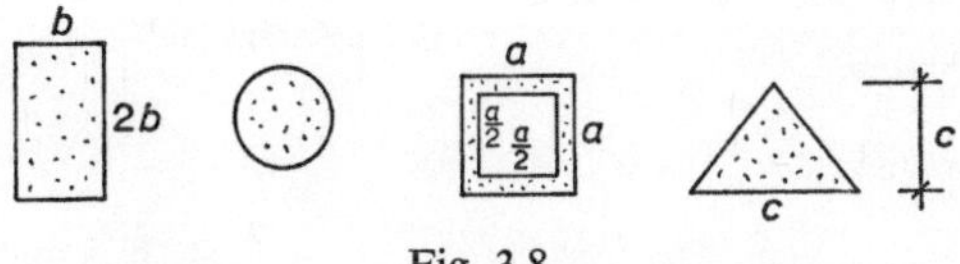

Fig. 3.8

(b) *Non-linearly elastic materials*

The 'standard' results of the preceding section apply only to the linear stress/strain relationship, which is the one commonly assumed for the working range of most construction materials. The assumption of plane sections remaining plane may be made even when the behaviour is non-linear, however, and thus the distribution of stress may be deduced. In the following example a material having elastic (reversible) but non-linear stress/strain properties is examined.

EXAMPLE 3.4

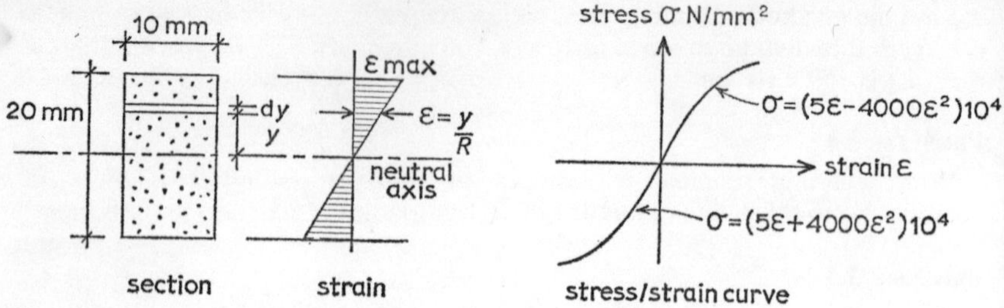

Fig. 3.9

We will determine the bending moment corresponding to a maximum stress of 15 N/mm² for this rectangular beam section, the material having the non-linear stress/strain relationship shown in the diagram. In view of the skew-symmetrical nature of the stress/strain curve we may conclude that the neutral axis will be at the centroid of the section and hence we may write,

$$M = 2\int_0^{10} (10\,\mathrm{d}y)\sigma \times y$$

$$= 20\int_0^{10} (5\varepsilon - 4\,000\varepsilon^2)10^4 y\,\mathrm{d}y$$

$$= 20\int_0^{10} \left(\frac{5y}{R} - 4\,000\,\frac{y^2}{R^2}\right) 10^4 y\,\mathrm{d}y$$

$$= 2\times 10^8\left(\frac{5}{3R} - \frac{10^4}{R^2}\right).$$

For a stress of 15 we may calculate the corresponding strain from,

$$15 = (5\varepsilon_{max} - 4\,000\varepsilon_{max}{}^2)10^4,$$

whence we find that $\varepsilon_{max} = 0.000\,5$, taking the lower solution of the quadratic equation.

Since $\varepsilon = \dfrac{y}{R}$ we may write, $0.000\,5 = \dfrac{10}{R}$, giving $R = 2\times 10^4$. Thus we find that

$$M = 2\times 10^8\left(\frac{5}{3\times 2\times 10^4} - \frac{10^4}{4\times 10^8}\right)$$

$$= 1.17\times 10^4 \text{ Nmm}.$$

(c) *Non-homogeneous sections*

Non-homogeneity of mechanical properties may result from the use of different materials to make up the member, or from the variation of the properties of the same material according to strain level, time, temperature, etc. Many of the latter will be non-elastic effects, such as the formation of 'plastic hinges' considered later.

EXAMPLE 3.5. *Composite section of linearly elastic materials*

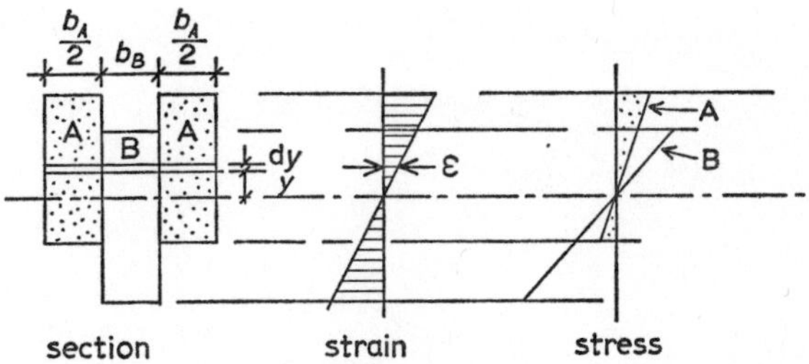

Fig. 3.10

The elastic moduli of the materials A and B are E_A and E_B respectively.

Making the assumption that the two materials are perfectly bonded together the strain ε at a level y above the neutral axis is the same in both,

$$\varepsilon = \frac{y}{R}.$$

Thus the stresses in the two materials are

$$\sigma_A = E_A \frac{y}{R}, \quad \text{and} \quad \sigma_B = E_B \frac{y}{R},$$

and the total longitudinal force on the layer thickness dy is

$$(E_A b_A + E_B b_B) \frac{y}{R} \, \mathrm{d}y.$$

Since there is no resultant longitudinal force on the section we have

$$\frac{1}{R} \int_{\text{section}} (E_A b_A + E_B b_B) y \, \mathrm{d}y = 0.$$

That is, the neutral axis passes through a 'centroid' which is found by 'weighting' the two materials according to their moduli of elasticity. Alternatively this may be thought of as replacing one material by an equivalent area of the other, for example,

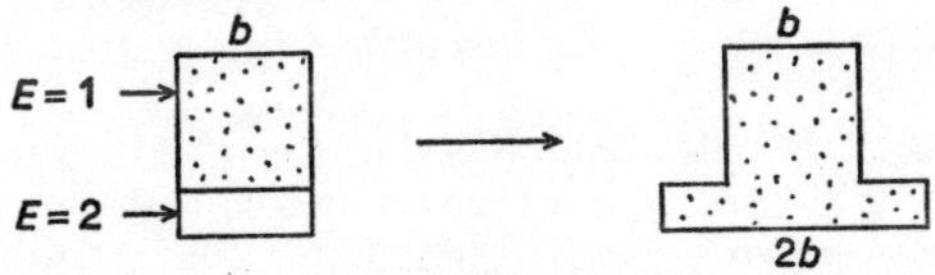

Fig. 3.11

The bending moment on the section is the couple resultant of the longitudinal forces;

$$M = \frac{1}{R}\int_{\text{section}} (E_A b_A + E_B b_B) y^2 dy,$$

$$= \frac{1}{R}(\;_A I_A + E_B I_B),$$

where I_A and I_B are the second moments of area of each material about the neutral axis of the section.

The stresses at level y are then

$$\sigma_A = E_A \frac{My}{(E_A I_A + E_B I_B)},$$

$$\sigma_B = E_B \frac{My}{(E_A I_A + E_B I_B)}.$$

Or, if the section is transformed by replacing material B with its equivalent of material A,

$$I' = \left(I_A + \frac{E_A}{E_B} I_B\right),$$

and

$$\sigma_A = \frac{My}{I'},$$

$$\sigma_B = \frac{E_B}{E_A}\frac{My}{I'}.$$

EXAMPLE 3.6. *Composite section of materials with variable properties: reinforced concrete*

Within the range of working stresses steel is approximately linearly elastic. Concrete, however, although roughly linearly elastic for short-term loading, tends to 'creep' significantly under sustained loading over a period of years. A realistic stress/strain curve for concrete is given in problem 3.9; however, the following analysis assuming linear behaviour and using a modified modulus of elasticity to take account of creep has been widely used in the design of reinforced concrete members.

The tensile strength of concrete is usually about one-tenth of the compressive strength, which means that the behaviour of the concrete below the neutral axis will be different from that above. The most common reason for reinforcing concrete in flexure is to make up for this lack of tensile strength as in the following example. However, reinforcement is frequently used to provide additional compressive strength, in both columns and beams, and further tensile reinforcement in the form of links or bent-up bars is used in regions where shear forces produce (diagonal) tensile stresses. These latter aspects are not considered in the following analysis, which is solely concerned with the reinforcement used to carry the direct tensile stresses of bending. The concrete, which is extensively, if finely, cracked below the neutral axis at working loads, is assumed to have no tensile strength whatever. Plane sections are assumed to remain plane and the steel and concrete are assumed to be perfectly bonded together.

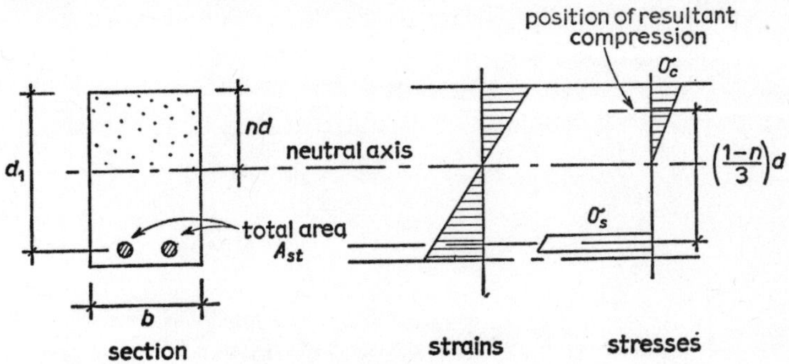

Fig. 3.12

Using the expression for strain derived in §3.1(a) and letting E_s and E_c be the Young's moduli for steel and concrete respectively

$$\sigma_c = \frac{E_c n d_1}{R} \quad \text{and} \quad \sigma_s = \frac{E_s(1-n)d_1}{R}. \tag{3.5}$$

Since there is no resultant force on the section

$$\frac{\sigma_c}{2} b n d_1 = \sigma_s A_{st}. \tag{3.6}$$

Equations (3.5) and (3.6) may be combined to give a quadratic equation for n,

$$\frac{b d_1}{A_{st}} \frac{E_c}{E_s} n^2 + n - 1 = 0. \tag{3.7}$$

The bending moment on the section, being equal to the couple formed by the concrete in compression and the steel in tension, may be expressed in terms of either, thus:

$$M \text{ (steel)} = \sigma_s A_{st}\left(1 - \frac{n}{3}\right) d_1$$

or (3.8)

$$M \text{ (concrete)} = \frac{\sigma_c}{2} b n d_1 \left(1 - \frac{n}{3}\right) d_1.$$

Equations (3.7) and (3.8) enable the stresses in both materials to be found.

(d) *Non-elastic materials*

When the material from which a member is made has a stress/strain curve approximating to that shown in Fig. 3.13, as is the case with most structural steels, the bending moment may be increased until virtually the whole section is at the yield stress, when a 'plastic hinge' is formed. At a plastic hinge, deformation will continue to increase under constant bending moment, and if a structure is loaded up until sufficient of these hinges are formed, collapse will occur.

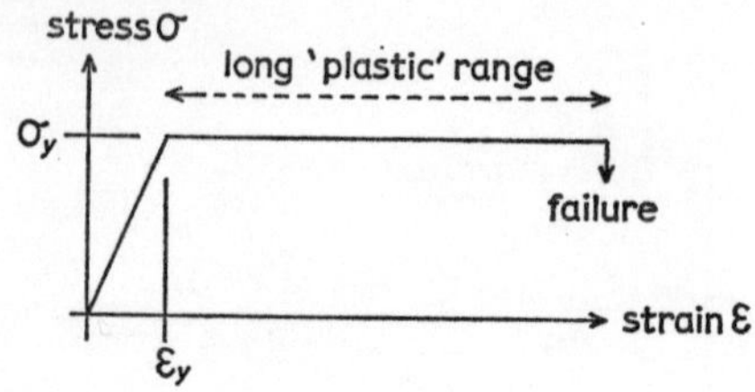

Fig. 3.13

Stress/strain relationship for an ideal elastic/plastic material in both tension and compression.

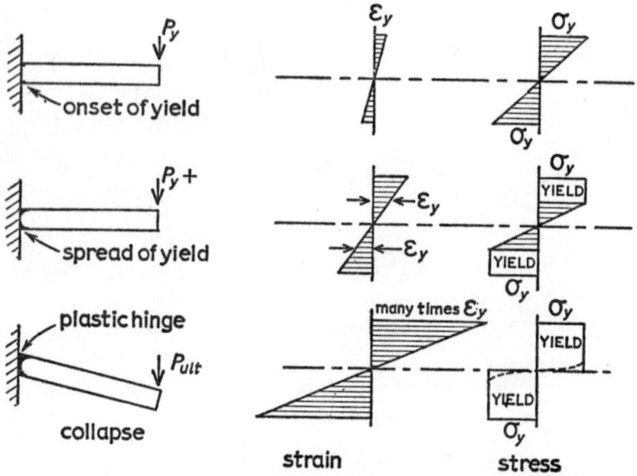

Fig. 3.14

Diagram showing the strain and stress distribution over a section as the bending moment is increased until a 'plastic hinge' is formed.

It is found that the assumption of plane sections remaining plane gives a satisfactory prediction of behaviour, even when the material of a member is so stressed beyond its elastic limit. It should be noted, of course, that a member bent thus far would not return to its original shape on the removal of load. Most practical structures behave substantially elastically within the range of working loads, but a consideration of their behaviour beyond this range is necessary for an assessment of the margin of safety against collapse.

The final stress distribution corresponds to the 'plastic hinge' condition, when the bending moment is denoted by M_p.

Referring to the section shown in Fig. 3.1 and using the stress diagram for the 'plastic hinge' condition above, the requirement that there is no resultant force on section gives

$$\int_{\text{section}} \sigma b dy = 0, \text{ where } \sigma = +\sigma_y \text{ above the neutral axis and } -\sigma_y \text{ below.}$$

Notice that this equation expresses the condition that the *neutral axis divides the section into equal areas.* The ultimate bending moment M_p is then simply found; for example, for a rectangular section, Fig. 3.15,

$$M_p = \left(\sigma_y b \frac{d}{2}\right)\frac{d}{2} = \sigma_y \frac{bd^2}{4}.$$

That is, the 'plastic' section modulus for a rectangular section is

$$Z_p = \frac{bd^2}{4}.$$

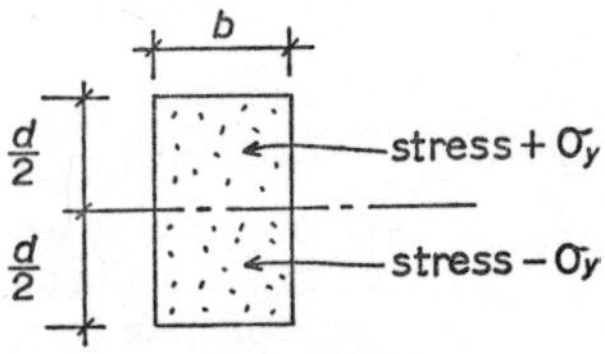

Fig. 3.15

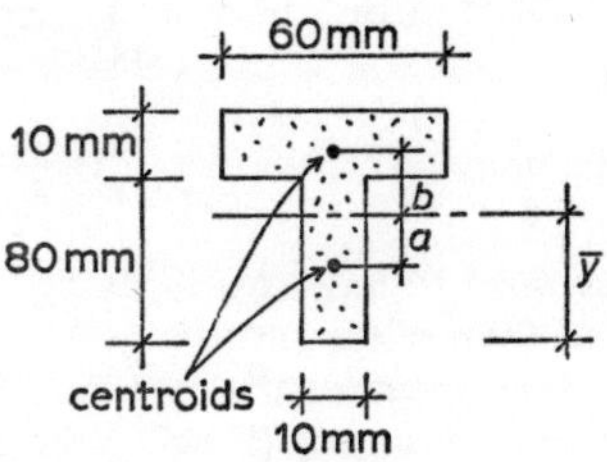

Fig. 3.16

The ratio between the elastic and plastic moduli for a section, $\frac{Z_p}{Z}$, is often referred to as the *shape factor*, and for common steel I-sections is about 1.15.

EXAMPLE 3.7

For a T-section, Fig. 3.16, whose yield stress is 100 N/mm², the total area is $(60+80)\times 10 = 1\,400$ mm².

Therefore $\bar{y} = \frac{1\,400}{2} \div 10 = 70$ mm,

$$a = \frac{70}{2} = 35 \text{ mm},$$

and $b = \frac{600\times 15 + 100 \times 5}{700} = 13.6$ mm, giving $$a+b = 48.6 \text{ mm.}$$

Thus $$M_p = 100 \times 70 \times 48.6 = 340 \text{ kN mm.}$$

PROBLEMS

PROBLEM 3.7

A beam section 250 mm deep is formed from a steel plate 12 mm thick sandwiched between two timber joists, each 50 mm thick. If the moduli of elasticity of the two materials are in the ratio of 20:1 and the limiting stresses are 200 and 8 N/mm² for the steel and timber respectively, what is the maximum bending moment to which the section may be subjected, assuming linearly elastic behaviour? (28.3 kNm)

PROBLEM 3.8

A rod is made up of fine filaments of high strength material embedded uniformly in a matrix of weaker material. The proportion of filaments in the cross section is 1% of the area and their modulus of elasticity is 200 times that of the matrix. What proportion of an applied bending moment is carried by the filaments, assuming linearly elastic behaviour? (66.7%)

PROBLEM 3.9

It has been found from tests on reinforced concrete beam sections that the strains at failure are approximately as shown in the diagram. Using the stress/strain curves given for the two materials, calculate the ultimate bending moment for the section shown (making the usual assumption for concrete in tension). The stress/strain curves are shown for concrete in compression and for steel in tension. (69.9 kNm)

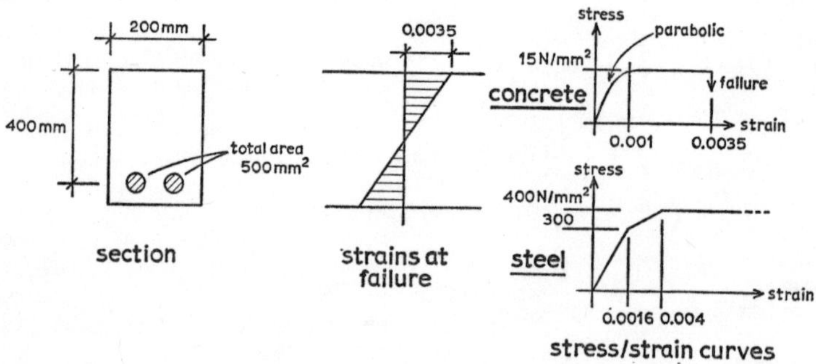

PROBLEM 3.10

Determine the ultimate bending moment which may be expected for the I-section shown in the diagram, the material having the relatively brittle nature indicated by the accompanying stress/strain curve. What will be the local radius of curvature at failure? (12.8 kNm, 15 m)

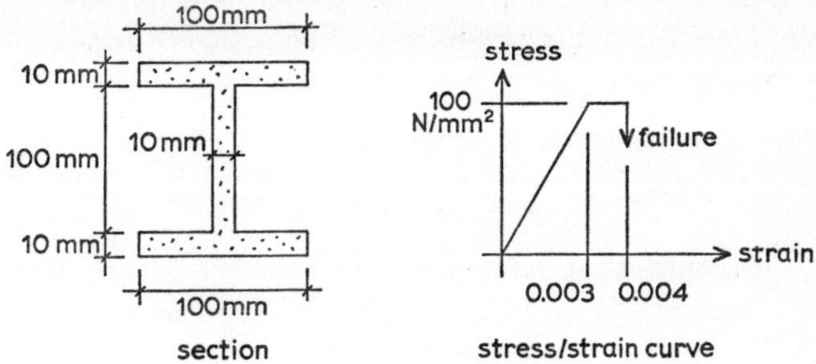

PROBLEM 3.11

The diagram shows the cross section of a 'sandwich panel' of aluminium and fibre board, the modulus of elasticity of the fibre board being one-tenth of that of the aluminium. Compare the bending moment on a unit width of panel with that which would cause the same maximum stress on the same thickness of solid aluminium, assuming linearly elastic behaviour. (1:1.36)

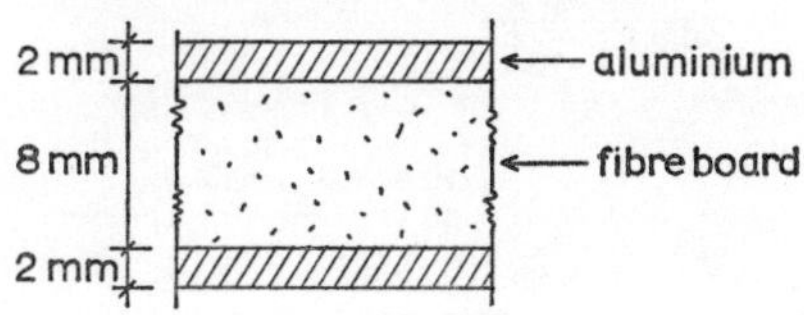

(e) *Non-symmetrical sections of linearly elastic material*

For the general section shown in Fig. 3.17, the applied bending moment has been resolved into components about the X and Y axes, neither of

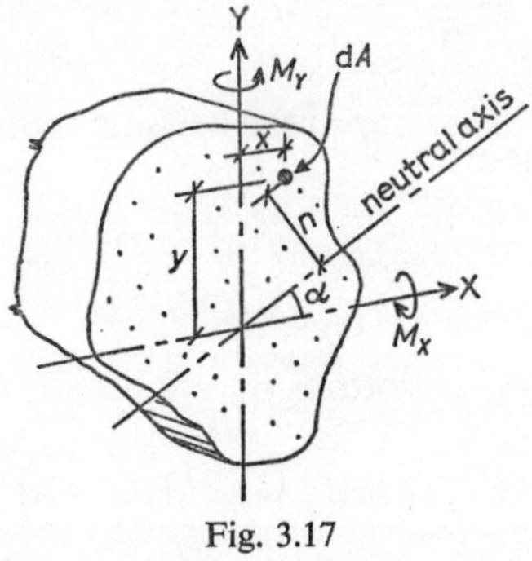

Fig. 3.17

which coincides with the neutral axis. As before, the assumption of plane cross sections means that strains, and hence for this material, stresses, are proportional to the distance from the neutral axis. It will also follow that the neutral axis passes through the centroid of the section.

The signs of the bending moment components are determined according to the right-hand screw rule with respect to the positive direction of the co-ordinate axes.

The (tensile) stress on the element $\mathrm{d}A$, $\sigma = \dfrac{En}{R}$, where

$$n = y\cos\alpha - x\sin\alpha.$$

Taking moments about the X-axis

$$M_x = \frac{E}{R}\int_{\text{section}} (y\cos\alpha - x\sin\alpha)\,\mathrm{d}Ay,$$

$$= \frac{E}{R}(I_{xx}\cos\alpha - I_{xy}\sin\alpha),$$

where I_{xx} is the second moment of area of the section about the X-axis, $\int y^2\mathrm{d}A$, and I_{xy} is the 'product of area' of the section about the axes, $\int xy\mathrm{d}A$. (See notes on p. 50.)

Similarly, $$M_y = \frac{E}{R}(I_{yy}\sin\alpha - I_{xy}\cos\alpha).$$

Thus $$\frac{M_x}{M_y} = \frac{I_{xx}\cos\alpha - I_{xy}\sin\alpha}{I_{yy}\sin\alpha - I_{xy}\cos\alpha},$$

and $$\tan\alpha = \frac{M_y I_{xx} + M_x I_{xy}}{M_x I_{yy} + M_y I_{xy}}. \tag{3.9}$$

The stress may now be evaluated

$$\sigma = \frac{E}{R}n = \frac{E}{R}(y\cos\alpha - x\sin\alpha),$$

and substitution for $\dfrac{E}{R}$ from one of the bending moment expressions above gives

$$\sigma = \frac{M_y}{I_{yy}\sin\alpha - I_{xy}\cos\alpha}(y\cos\alpha - x\sin\alpha).$$

Dividing through top and bottom by $\cos\alpha$ and substituting from (3.9) gives

$$\sigma = \frac{(M_x I_{yy} + M_y I_{xy})y - (M_y I_{xx} + M_x I_{xy})x}{I_{xx}I_{yy} - I_{xy}^{\,2}}. \tag{3.10}$$

If we choose the axes X and Y such that $I_{xy} = 0$, then I_{xx} and I_{yy} take on maximum and minimum values respectively and the axes are referred to as '*principal axes*' (see below). The equations (3.9) and (3.10) then reduce to the simpler forms:

$$\tan\alpha = \frac{M_y I_{xx}}{M_x I_{yy}} \tag{3.11}$$

and
$$\sigma = \frac{M_x}{I_{xx}}y - \frac{M_y}{I_{yy}}x. \tag{3.12}$$

Principal axes

Considering a general pair of axes X', Y' inclined at β to X, Y, the co-ordinate transformations are

$$x' = x\cos\beta + y\sin\beta$$
$$y' = y\cos\beta - x\sin\beta$$

hence
$$I_{y'y'} = \int_{\text{section}} \mathrm{d}A(x\cos\beta + y\sin\beta)^2,$$
$$= I_{yy}\cos^2\beta + I_{xx}\sin^2\beta + 2I_{xy}\cos\beta\sin\beta.$$

Similarly,
$$I_{x'x'} = I_{xx}\cos^2\beta + I_{yy}\sin^2\beta - 2I_{xy}\cos\beta\sin\beta,$$

and
$$I_{x'y'} = I_{xx}\cos\beta\sin\beta - I_{yy}\cos\beta\sin\beta + I_{xy}(\cos^2\beta - \sin^2\beta). \tag{3.13}$$

Adding the first two above gives the following:

$$I_{x'x'} + I_{y'y'} = I_{xx} + I_{yy}.$$

Therefore maximum $I_{x'x'}$ corresponds to minimum $I_{y'y'}$ and vice versa.

For maximum $I_{x'x'}$,

$$\frac{d}{\mathrm{d}\beta}(I_{x'x'}) = 0,$$

that is,
$$-2I_{xx}\cos\beta\sin\beta + 2I_{yy}\sin\beta\cos\beta - 2I_{xy}\cos 2\beta = 0.$$

Comparing this with the third of equations (3.13) it can be seen that extreme values of $I_{x'x'}$ and $I_{y'y'}$ occur when $I_{x'y'} = 0$, when

$$\tan 2\beta = \frac{2I_{xy}}{I_{yy} - I_{xx}}.$$

The symmetrical sections considered previously constitute a special case, because $I_{xy} = 0$ if either X or Y is an axis of symmetry. For a single bending moment component M_x equation (3.11) confirms that the neutral axis coincides with the X-axis and equation (3.12) reduces to the form derived in § 3.1 (a).

Comparing the equations (3.13) with those for the transformation of plane stresses or plane strains in the theory of elasticity,* it will be seen that the form is identical. Quantities for which this type of transformation applies are known as *tensors*. The graphical construction usually associated with plane stress transformations, Mohr's circle, may therefore be used for second moments of area.

NOTES ON 'PRODUCTS OF AREA'

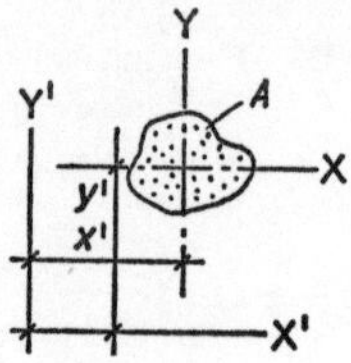

Fig. 3.18

The parallel axes theorem applies to products of area, that is

$$I_{x'y'} = I_{xy} + Ax'y'$$

The 'three point' substitution for triangles is also applicable. Note that I_{xy} is zero if X or Y are axes of symmetry.

EXAMPLE 3.8

Given that I_{xx}, I_{yy}, I_{xy} are 12, 6, 4 respectively, locate the principal axes of the section, and evaluate $I_{x'x'}$, $I_{y'y'}$, $I_{x'y'}$, where X', Y' are inclined at $+30°$ to the axes X, Y.

By calculation:

For a principal axis:

$$\tan 2\beta = \frac{2I_{xy}}{I_{yy}-I_{xx}} = \frac{-8}{12-6} = -1.333,$$

whence the least value of β is 63° 25′.

For the axes X', Y',

$$I_{x'x'} = 12\times\tfrac{3}{4}+6\times\tfrac{1}{4}-4\times\frac{\sqrt{3}}{2} \quad = 7.04$$

$$I_{y'y'} = 12\times\tfrac{1}{4}+6\times\tfrac{3}{4}+4\times\frac{\sqrt{3}}{2} \quad = 10.96$$

$$I_{x'y'} = 12\times\frac{\sqrt{3}}{4}-6\times\frac{\sqrt{3}}{4}+4\times\tfrac{1}{2} = 4.60.$$

* See *Stress and Strain*, T. R. Graves-Smith (Chatto and Windus).

Graphical solution
(Mohr's circle)

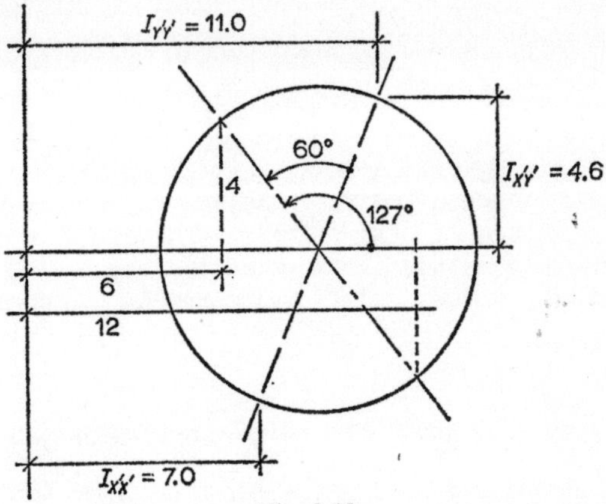

Fig. 3.19

EXAMPLE 3.9

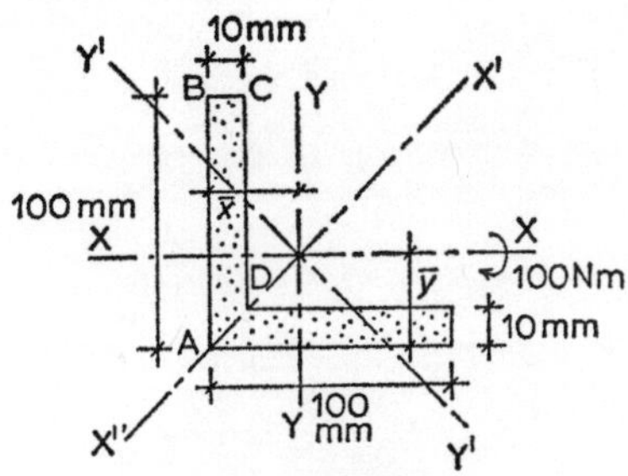

Fig. 3.20

The problem is to find the maximum stress produced in the equal angle section shown, by a bending moment of 100 Nm about the X-X axis.

Evaluation of section properties:

$$\bar{x} = \bar{y} = \frac{100 \times 10 \times 50 + 90 \times 10 \times 5}{1\,900} = 28.7 \text{ mm}$$

$$I_{xx} = I_{yy} = \left(\frac{10 \times 100^3}{12} + 10 \times 100 \times 21.3^2\right) + \left(\frac{90 \times 10^3}{12} + 90 \times 10 \times 23.7^2\right)$$
$$= 1.8 \times 10^6 \text{ mm}^4$$

$$I_{xy} = 1\,000 \times (-23.7) \times 21.3 + 900 \times 26.3 \times (-23.7) = -1.06 \times 10^6 \text{ mm}^4.$$

Evaluating coefficients in equation (3.10),

$$\frac{(M_x I_{yy} + M_y I_{xy})}{(I_{xx} I_{yy} - I_{xy}{}^2)} = 0.0851$$

$$\frac{(M_y I_{xx} + M_x I_{xy})}{(I_{xx} I_{yy} - I_{xy}{}^2)} = -0.506.$$

Therefore the stress at A

$$= 0.085\,1\times(-2.87)-(-0.050\,6)\times(-28.7) = -3.92$$

the stress at B

$$= 0.085\,1\times 71.3-(-0.050\,6)\times(-28.7) \quad = \quad 4.59$$

and the stress at C

$$= 0.085\,1\times 71.3-(-0.050\,6)\times(-18.7) \quad = \quad 5.10.$$

That is, the maximum stress is 5.10 N/mm² (tension) at C.

An alternative approach would be to evaluate the section properties with respect to the *principal* axes X' and Y' (principal because X' is an axis of symmetry). The simpler formulae, (3.13), could then be applied, the bending moment being resolved into X' and Y' components for this purpose.

PROBLEMS

PROBLEM 3.12

Evaluate I_{xx}, I_{yy}, I_{xy} and locate principal axes for the following sections:

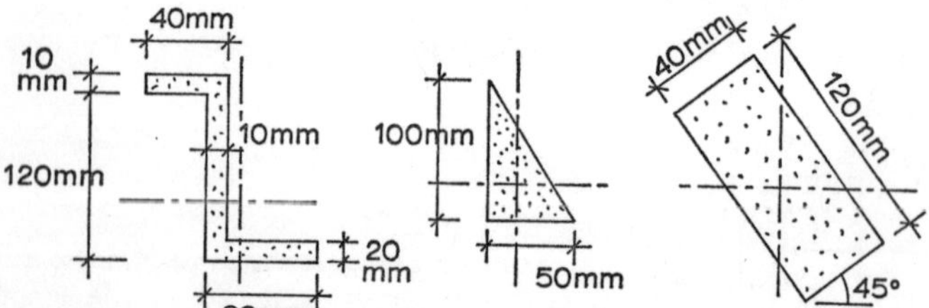

545×10^3	139×10^3	320×10^3
104×10^3	34.8×10^3	320×10^3
-167×10^3	-4.2×10^3	-320×10^3
18.5°	2.25°	45°

PROBLEM 3.13

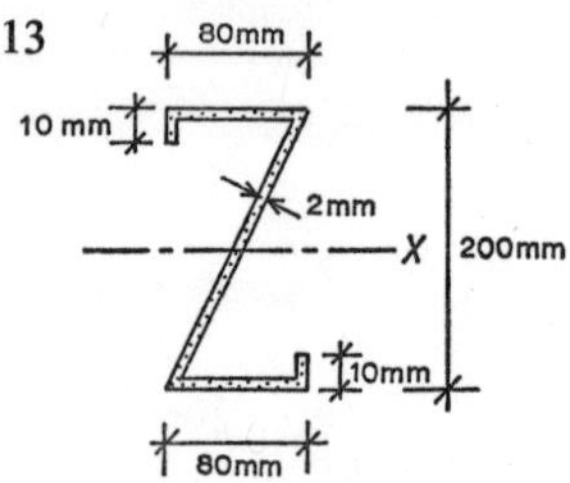

Determine the maximum stress which occurs when the section shown is subjected to a bending moment of 3 000 kNmm about the X-axis.

(90 N/mm²)

PROBLEM 3.14

The box section is subjected to bending moments about the X and Y-axes as shown. What angle does the neutral axis make with the X-axis?

(58.75°)

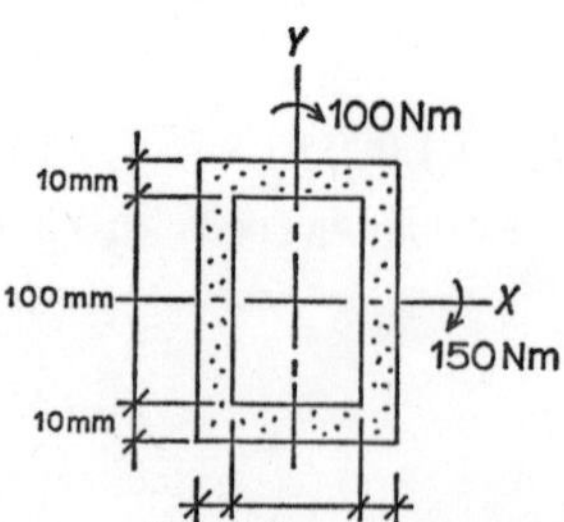

3.2 Shear stresses

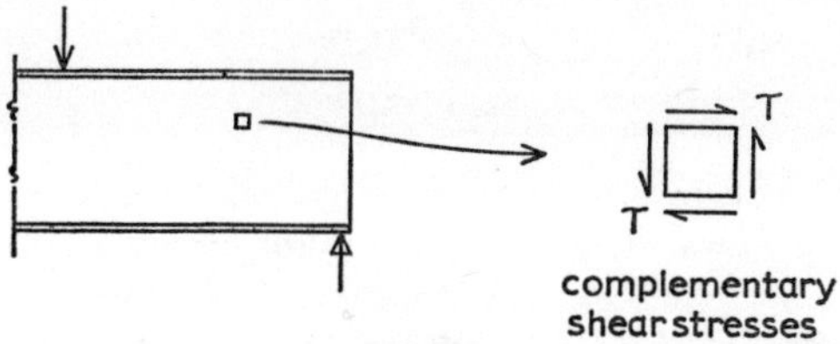

Fig. 3.21

The distribution of shear stress over a symmetrical cross section of homogeneous linearly elastic material is examined by considering the complementary shear stresses between the longitudinal 'layers' of the member. The thin slice of member shown in Fig. 3.22 is from a region where the bending moment is varying; that is, a region where the shear force is not zero.

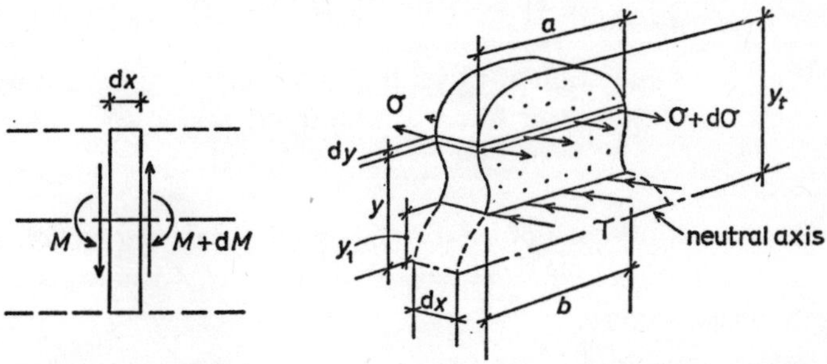

Bending moments and shear forces acting on a thin slice

View of slice showing direct stresses at level y and shear (longitudinal) stress at level y

Fig. 3.22

The resultant longitudinal force on the ends of the layer thickness dy

$$= (\sigma + \mathrm{d}\sigma)a\mathrm{d}y - \sigma a \mathrm{d}y$$

$$= \mathrm{d}\sigma a \mathrm{d}y.$$

Thus the net longitudinal force on the faces of the part of the slice which is above y_1

$$= \int_{y_1}^{y_t} \mathrm{d}\sigma a \mathrm{d}y.$$

Now, $\sigma = \dfrac{My}{I}$, so that $\mathrm{d}\sigma = \dfrac{y\mathrm{d}M}{I}$, and the resultant force $$= \frac{dM}{I}\int_{y_1}^{y_t} ay\,dy,$$

$$= \frac{dM}{I} A\bar{y},$$

where $A\bar{y}$ is first moment of area of section above y_1 about the neutral axis.

This force is balanced by the horizontal shear stress τ between the upper and lower parts of the slice at level y_1.

Thus $$\tau = \frac{\frac{dM}{I} A\bar{y}}{b dx},$$

$$= \frac{dM}{dx}\frac{A\bar{y}}{bI}.$$

But $$\frac{dM}{dx} = Q, \text{ the shear force, so that } \tau = \frac{QA\bar{y}}{bI}. \qquad (3.14)$$

By definition the summation of τ over the whole section at a point will give the stress resultant (the shear force) at that point.

EXAMPLE 3.10

Investigate the distribution of shear stress over a rectangular section and an I-section at a point where the shear force is 10^5N:

(a) Rectangular section

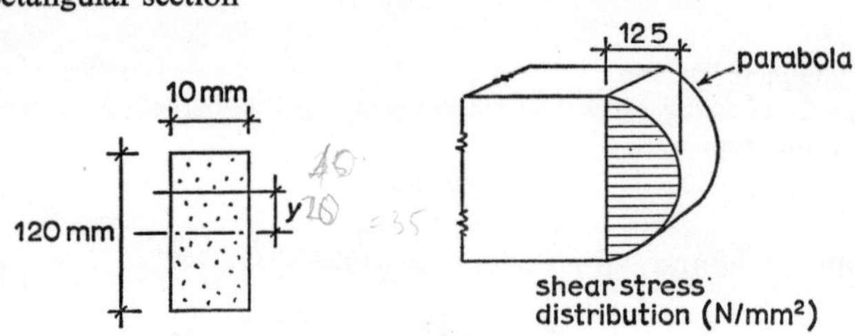

Fig. 3.23

At y above the neutral axis,

$$\tau_{(y)} = \frac{10^5 \times 10 \times (60-y) \times \frac{(60+y)}{2}}{10 \times 144 \times 10^4}$$

$$= \frac{5}{144}(3\,600 - y^2)$$

and maximum τ = 125 N/mm².

Note that

$$\int_{\text{section}} \tau b dy = \int_{-60}^{+60} \frac{5}{144}(3\,600 - y^2) \times 10\, dy = 10^5 \text{ (shear force).}$$

(b) I-section

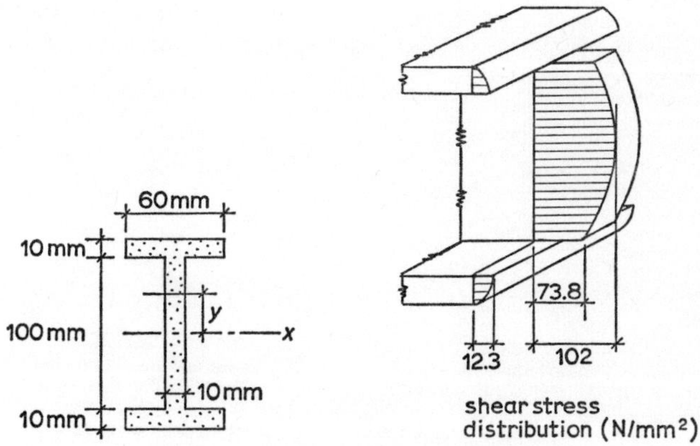

Fig. 3.24 (a)

$$I_{xx} = 2\left(\frac{60 \times 10^3}{12} + 60 \times 55^2\right) + \frac{10 \times 100^3}{12} = 447 \times 10^4.$$

In the web,

$$\tau_{(y)} = \frac{10^5\left(600 \times 55 + 10 \times (50 - y) \times \dfrac{(50 + y)}{2}\right)}{10 \times 447 \times 10^4}.$$

In the flanges,

$$\tau_{(y)} = \frac{10^5\left(60 \times (60 - y) \times \dfrac{(60 + y)}{2}\right)}{10 \times 447 \times 10^4}.$$

Hence the common assumption that the web of an I-beam 'takes all the shear'. In other words, the average shear stress is given approximately by

$$\tau = \frac{\text{shear force}}{\text{web area}}.$$

In this example this formula would give an average shear stress

$$\tau = \frac{10^5}{120 \times 10} = 83.3 \text{ N/mm}^2.$$

The distribution of (vertical) shear stress indicated in Fig. 3.24 (a) shows up clearly the inadequacy of this 'simple' theory in regions where there is a sudden change of dimension, in this case at the junction of web and flange.

Obviously there cannot be such a sudden drop in the intensity of shear stress at this point and in fact the stress would fall gradually from the web value to zero at the outside of the section, the zone of vertical shear stress effectively spreading at about 45° as in Fig. 3.24 (b). This means that the outer parts of the flanges would be substantially free of vertical shear stress. A further point is that an analysis similar to that used for the vertical shear stresses could be applied in the flanges to disclose the existence of horizontal shear stresses as indicated in Fig. 3.24 (c).

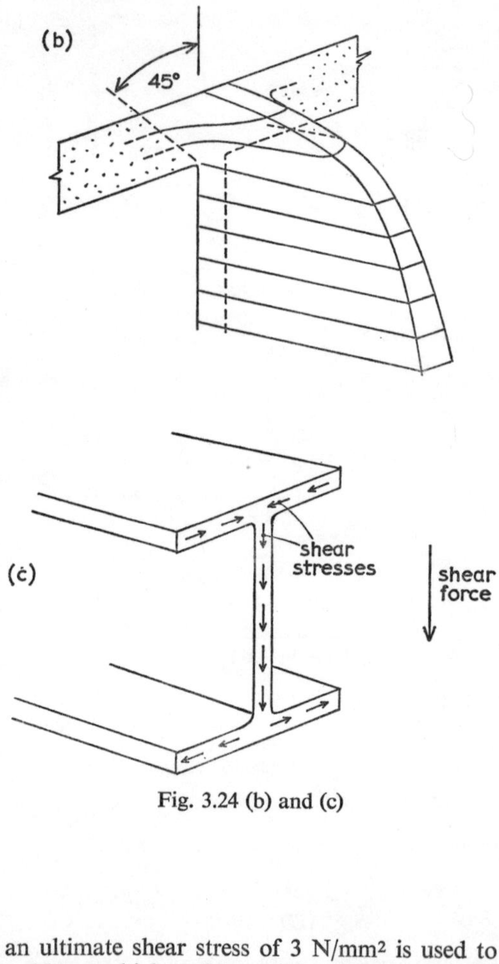

Fig. 3.24 (b) and (c)

PROBLEMS

PROBLEM 3.15

Glue having an ultimate shear stress of 3 N/mm² is used to bond together planks 100 mm × 10 mm thick to form a joist of 200 mm depth and 2 m span. At what value of uniformly distributed load would the laminations start to separate? (40 kN/m)

PROBLEM 3.16

The joist section shown below left is subjected to a shear force of 100 kN together with a bending moment of 100 kNm. What is the maximum principal stress in the web (a) at the neutral axis and (b) at the 'root' of the flange? (8.6, 18.6 N/mm²)

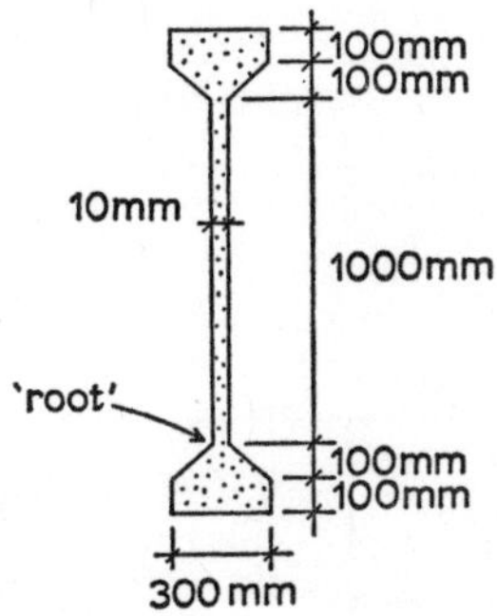

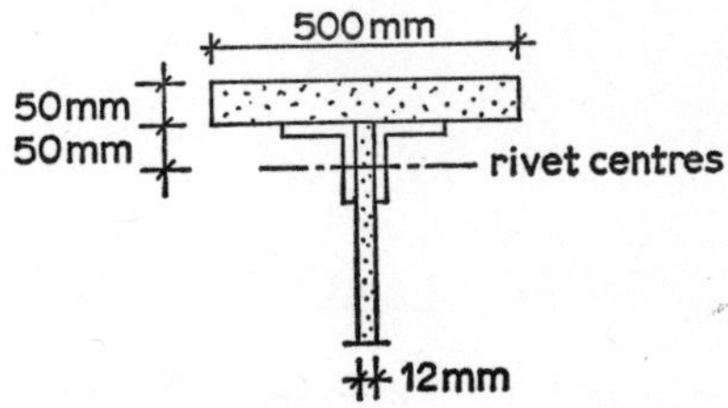

PROBLEM 3.17

The flanges of a steel I-girder, above right, are each 500 mm × 50 mm and are fixed to the 1 250 mm × 12 mm web with two angles by means of rivets which are capable of withstanding a shearing force of 25 kN at each shear plane. Determine approximately the spacing for the rivets at a point on the girder where the shear force is 750 kN. Ignore the areas of the angles and the areas lost for rivet holes. (93 mm)

3.3 Combination of stresses

Elastic bending stresses, being linear functions of loading, may be combined with other 'linear' stress systems by simple addition (super position).

EXAMPLE 3.11. *Brick wall subjected to lateral loading*

Fig. 3.25 shows a section through a brick wall and its foundation, the direct stress resultants at foundation level being a bending moment from the lateral loading and a vertical force from the weight of the wall. In the following analysis of the (direct) stresses at foundation level a metre length of wall is considered.

Maximum bending stress

$$\sigma_B = \frac{70 \times 2 \times 10^3 \times 6}{1\,000 \times 200^2} = 0.021 \text{ N/mm}^2.$$

Maximum gravity stress

$$\sigma_G = \frac{20 \times 10^3}{10^9} \times 2\,000 = 0.04 \text{ N/mm}^2.$$

The stress distribution is shown in (b). (Brickwork is generally assumed to have no tensile strength across the courses, and so this stress distribution is 'safe'.)

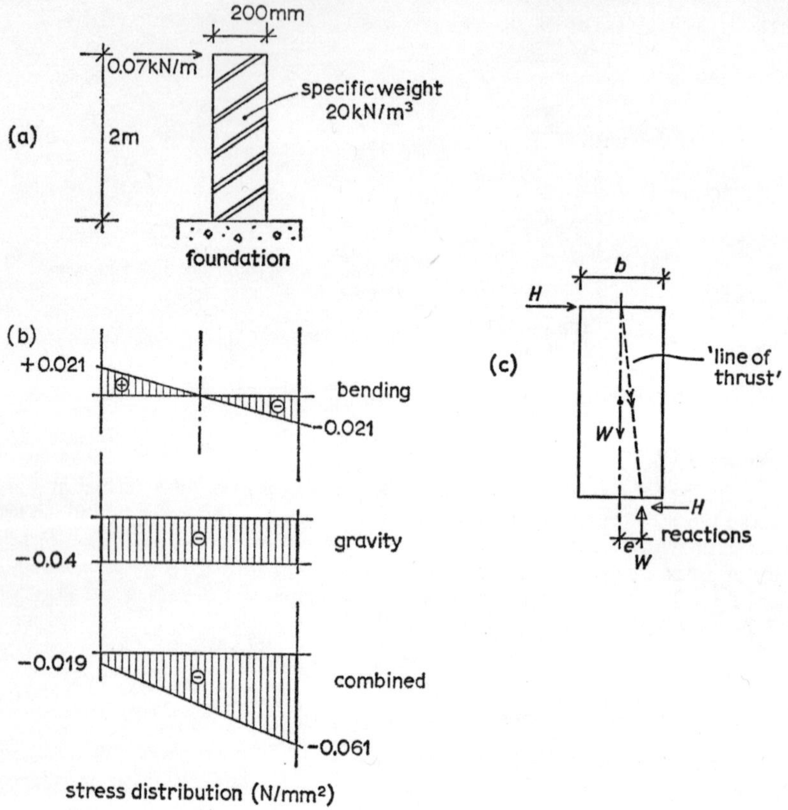

Fig. 3.25

Alternatively, the problem may be approached as follows:

The horizontal and vertical forces acting on a unit length of wall may be combined to a single resultant along the 'line of thrust' as shown in (c). The reactions from the foundation act where this line of thrust cuts the base, W being the normal component and H, the shear. The bending moment on the base can be now expressed as $W \times e$ (eccentricity) and the maximum and minimum stresses are $\sigma = \dfrac{W}{b} \pm \dfrac{6We}{b^2}$.

From this it can be seen that tension does not develop provided $e \ngtr \dfrac{b}{6}$ or *the line of thrust cuts the base inside the middle* $\frac{1}{3}$.

EXAMPLE 3.12. *Prestressed concrete*

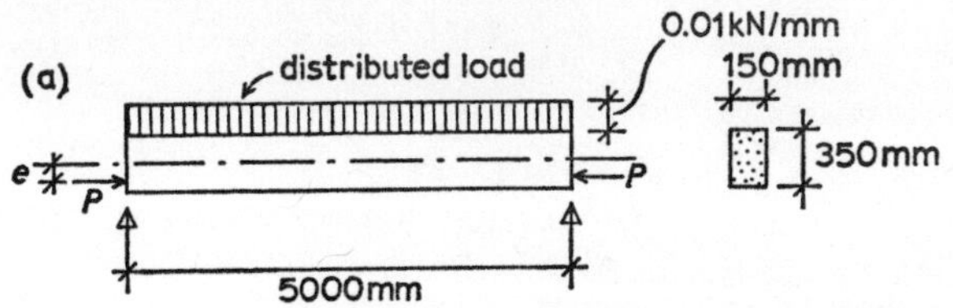

$$A = 150 \times 350 = 52\,500$$

$$Z = \frac{150 \times 350^2}{6} = 3\,060 \times 10^3$$

Fig. 3.26 (a)

P is the prestressing force and e its eccentricity.

The object of the prestressing is to set up compressive stresses in areas in which the imposed loading would cause tensile stresses (which the concrete could not otherwise sustain).

The problem is to find the least value of the prestressing force P and the corresponding value of the eccentricity e for the condition that the concrete stress must lie between zero and 0.010 5 kN/mm² (compressive) for both maximum and minimum bending moments.*

For the minimum (zero) bending moment condition it follows from the previous example that

$$e \ngtr \frac{350}{6}. \tag{3.15}$$

Also the maximum compressive stress (at the bottom of the section) must not exceed 0.010 5, that is

$$\frac{P}{52\,500} + \frac{Pe}{3\,060 \times 10^3} \ngtr 0.010\,5. \tag{3.16}$$

These conditions are shown in (b).

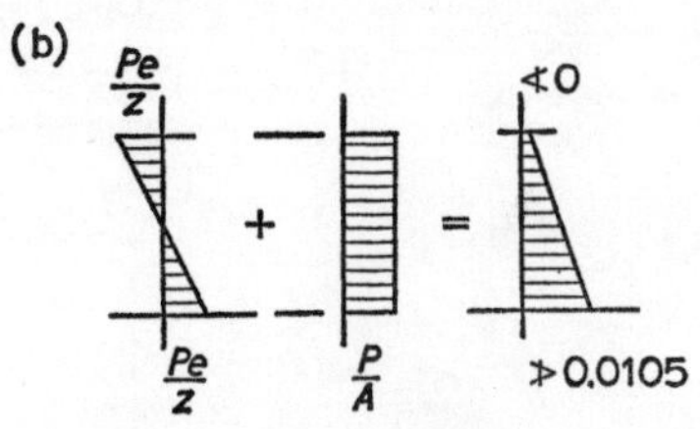

Fig. 3.26 (b)

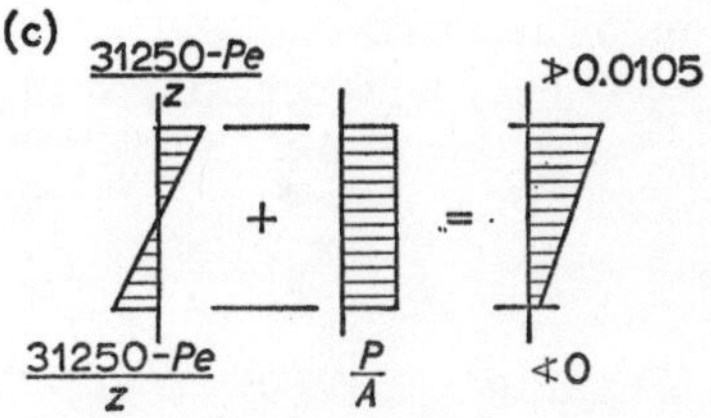

Fig. 3.26 (c)

For the maximum (negative) bending moment condition

$$\text{maximum} \quad M = \frac{0.01 \times 5\,000^2}{8} - Pe,$$

$$= 31\,250 - Pe.$$

* The minimum bending moment will occur when there is no loading on the beam, or at the ends of the beam when the loading is applied.

At the top of the beam,

$$\frac{31\,250-Pe}{3\,060\times10^3}+\frac{P}{52\,500}\ngtr 0.010\,5. \tag{3.17}$$

At the bottom of the beam,

$$\frac{-31\,250+Pe}{3\,060\times10^3}+\frac{P}{52\,500}\nless 0. \tag{3.18}$$

These conditions are shown in (c).

It is found that the least value of P is given by taking (3.15) and (3.16) as the limiting criteria, as can be seen in the interaction diagram Fig. 3.26 (d).

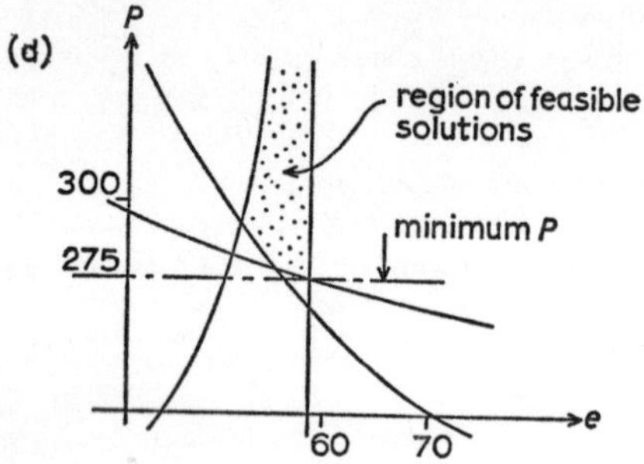

Interaction diagram

Fig. 3.26 (d)

The solution giving the minimum value of P is:

$$P = 275 \text{ kN}$$
$$e = 58.3 \text{ mm}$$

Problems

Problem 3.18

The diagram, below left, shows a section through a masonry dam with water on the vertical face.

Find the least value for d and the corresponding maximum base pressure for there to be no tension at the base of the dam. The specific weights of water and masonry may be taken as 10 and 20 kN/m³ respectively. (8.95 m)

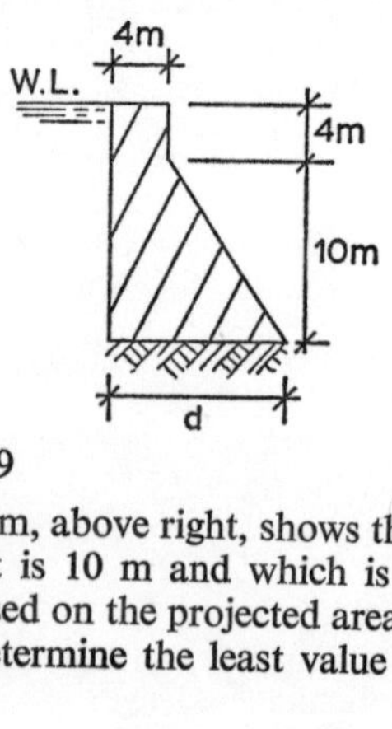

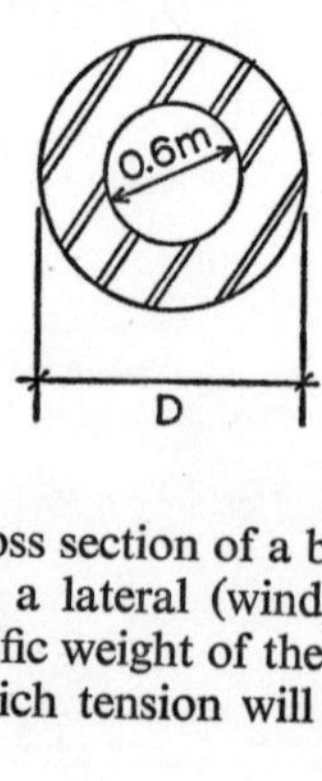

Problem 3.19

The diagram, above right, shows the annular cross section of a brick chimney whose height is 10 m and which is subjected to a lateral (wind) pressure of 1 kN/m² (based on the projected area). If the specific weight of the brickwork is 20 kN/m³ determine the least value of D for which tension will not develop. (1.71 m)

PROBLEM 3.20

The diagram, below left, shows the section of a prestressed concrete beam. Determine the least prestressing force required and the corresponding eccentricity for the following conditions:

Concrete stresses (compression)

maximum	20 N/mm^2
minimum	0.

Bending moments (negative)

maximum	$2\,000 \times 10^6$ Nmm
minimum	600×10^6 Nmm.

(3 050 kN, 227 mm)

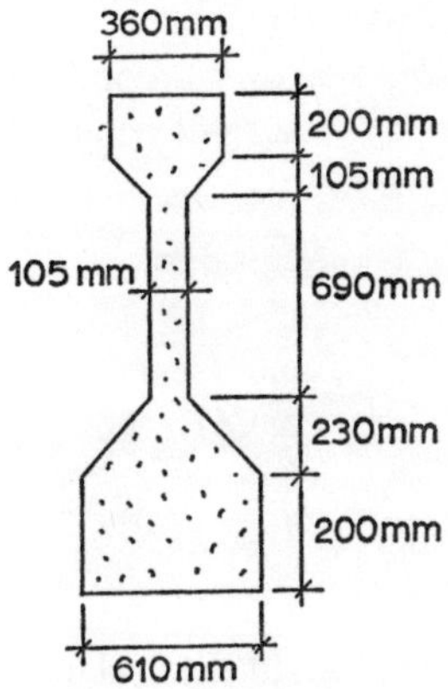

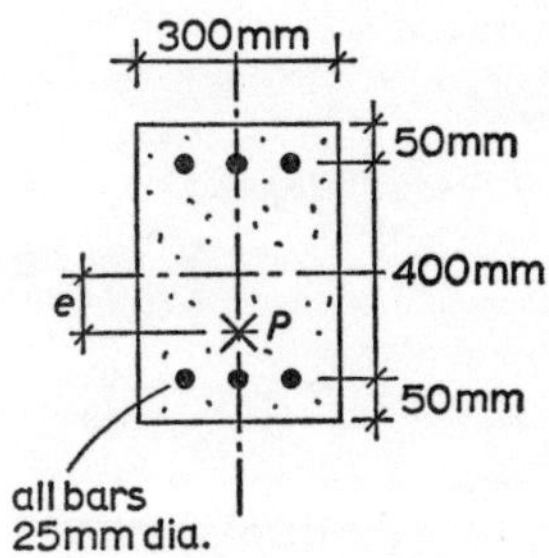

PROBLEM 3.21

The reinforced concrete column section shown, above right, carries an eccentric axial compressive load P. Find the value of e for which tension will just develop under a load $P = 1\,000$ kN.

What is the maximum compressive stress in the concrete when the eccentricity e is increased by 50 mm, the load P staying the same? The 'modular ratio' $\dfrac{E_s}{E_c}$ is to be taken as 15 and concrete must be assumed to be incapable of sustaining tensile stresses. (100 mm, 13.3 N/mm^2)

4

ELASTIC DISPLACEMENTS

For the uniplanar bending of a homogeneous member of linearly elastic material it was shown on page 36 that the relation between the applied bending moment and the radius of curvature of the member is $M = \frac{EI}{R}$ (equation (3.2)), where E is Young's modulus and I the second moment of area of the section. Thus the quantity EI is a measure of bending stiffness and is often referred to as the *flexural rigidity*.

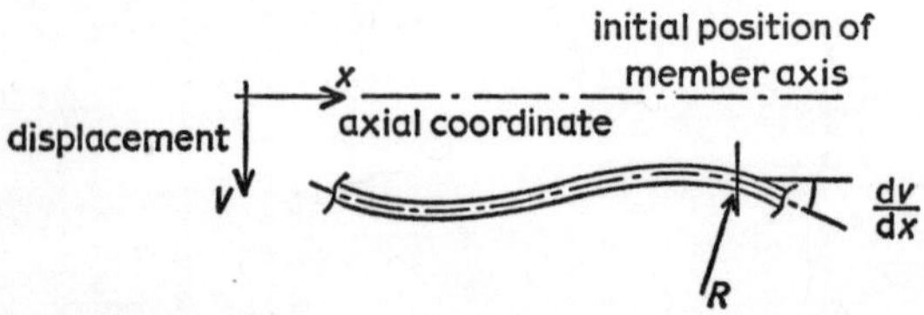

Fig. 4.1

Fig. 4.1 shows the displacement of an initially straight flexural member, the equation of the displaced axis, with respect to the initial position being $v = f(x)$.

The curvature at any point is given by

$$\frac{1}{R} = \frac{\dfrac{d^2v}{dx^2}}{\left[1-\left(\dfrac{dv}{dx}\right)^2\right]^{\frac{3}{2}}}. \tag{4.1}$$

For small displacements (in practice, displacements are usually limited to about $\frac{1}{300}\times$ span for simple beams), $\left(\frac{dv}{dx}\right)^2$ is *very* small compared with unity and the expression may be closely approximated by, $\frac{1}{R} = \frac{d^2v}{dx^2}$.

Substituting from (3.2) we obtain the (differential) equation of flexure for small displacements of an initially straight member

$$\frac{M}{EI} = \frac{d^2v}{dx^2}. \tag{4.2}$$

A similar expression may be used for the small displacements of initially *curved members* provided the radii of curvature are large compared with the lateral dimensions. In this case the *change of curvature* rather than curvature would become the subject of equation (3.2) and v would represent the displacement normal to the original longitudinal axis.

For a member of initial curvature $\frac{1}{R_0}$ we would obtain the following results

$$\frac{M}{EI} = \left(\frac{1}{R} - \frac{1}{R_0}\right)$$

$$= \frac{\mathrm{d}^2 v}{\mathrm{d}s^2},$$

where s is the distance along the curve from the origin.

For initially straight members, by combining equation (4.2) with equations (2.2) and (2.3) we obtain the following sequence of relations for the displacement and its derivatives:

$$\text{displacement} = v$$

$$\text{slope} = \frac{\mathrm{d}v}{dx}$$

$$\frac{\text{bending moment}}{\text{flexural rigidity}} = \frac{M}{EI} = \frac{\mathrm{d}^2 v}{\mathrm{d}x^2}$$

$$\frac{\text{shear force}}{\text{flexural rigidity}} = \frac{Q}{EI} = \frac{\mathrm{d}^3 v}{\mathrm{d}x^3}$$

$$\frac{\text{intensity of loading}}{\text{flexural rigidity}} = \frac{\omega}{EI} = \frac{\mathrm{d}^4 v}{\mathrm{d}x^4}.$$

Provided M, Q, ω and EI are analytic functions of x only (or constant), the slope and displacement equations may be found by direct integration of the loading, shear force or bending moment equations.

4.1 Direct solution of the differential equation of flexure

If the functions $\frac{\omega}{EI}$, $\frac{Q}{EI}$ and $\frac{M}{EI}$ are all continuous, we may start at whichever one is convenient and proceed to determine the slope and displacement directly by a series of integrations to which appropriate boundary conditions are applied.

EXAMPLE 4.1

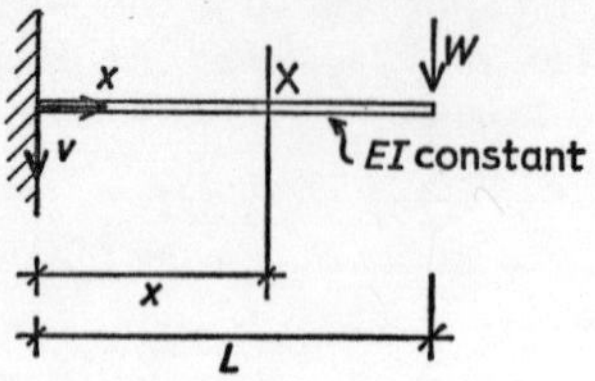

Fig. 4.2

The expression for bending moment at the general position X is

$$M = W(L-x).$$

We may therefore write $$EI\frac{d^2v}{dx^2} = W(L-x).$$

Integrating, $$EI\frac{dv}{dx} = \frac{-W}{2}(L-x)^2+C_1$$

and $$EIv = \frac{W}{6}(L-x)^3+C_1x+C_2,$$

where C_1 and C_2 are constants of integration.

In order to find the values of C_1 and C_2 it is necessary to take into account the boundary conditions of the problem.

When $x = 0$, the slope $\frac{dv}{dx} = 0$.

Therefore, $$0 = -\frac{WL^2}{2}+C_1, \text{ giving } C_1 = \frac{WL^2}{2}.$$

When $x = 0$, the displacement $v = 0$.

Therefore $$0 = \frac{WL^3}{6}+C_2, \text{ giving } C_2 = -\frac{WL^3}{6}.$$

The displacement equation then becomes:

$$EIv = \frac{W}{6}(L-x)^3+\frac{W}{2}L^2x-\frac{W}{6}L^3.$$

The displacement at the free end of the cantilever is found to be $\frac{WL^3}{3EI}$ and the slope, $\frac{WL^2}{2EI}$.

If we put in practical values we can get some idea of the order of magnitude of these quantities. A steel I-section, 203 mm × 133 mm, used as a cantilever 3 m long with an end load of 10 kN will displace 19.1 mm and rotate 0.009 6 radians at the free end. These obviously fall well within the 'small deflection' range for which it is justifiable to replace $\left(1-\left(\frac{dv}{dx}\right)^2\right)$ by 1.

EXAMPLE 4.2

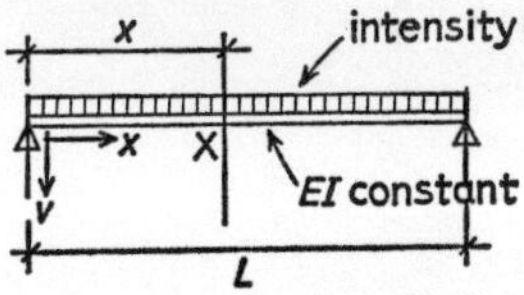

Fig. 4.3

At the general position X we may write

$$EI\frac{d^4v}{dx^4} = \omega.$$

Integrating:

(shear force) $$EI\frac{d^3v}{dx^3} = \omega x + C_1$$

(bending moment) $$EI\frac{d^2v}{dx^2} = \frac{\omega x^2}{2} + C_1x + C_2$$

$$EI\frac{dv}{dx} = \frac{\omega x^3}{6} + C_1\frac{x^2}{2} + C_2x + C_3$$

$$EIv = \frac{\omega x^4}{24} + \frac{C_1x^3}{6} + \frac{C_2x^2}{2} + C_3x + C_4$$

where C_1, C_2, C_3 and C_4 are constants of integration.

The boundary conditions are that bending moment and displacement are both zero at $x = 0$ and $x = L$.

Hence we obtain

$$C_1 = -\frac{\omega L}{2},\ C_2 = 0,\ C_3 = \frac{\omega L^3}{24} \text{ and } C_4 = 0.$$

The displacement equation is then:

$$EIv = \frac{\omega x^4}{24} - \frac{\omega Lx^3}{12} + \frac{\omega L^3x}{24}.$$

The maximum displacement obviously occurs at $x = \frac{L}{2}$ and is equal to $\frac{5\omega L^4}{384EI}$.

EXAMPLE 4.3

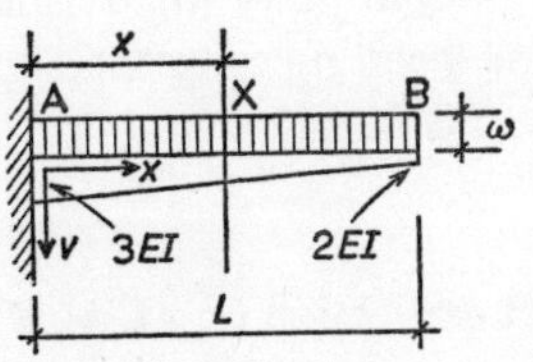

Fig. 4.4

For the member shown in Fig. 4.4 the flexural rigidity varies linearly from $3EI$ to $2EI$.

At X the bending moment $= \frac{\omega}{2}(L-x)^2$

and the flexural rigidity $= 3EI - \frac{EIx}{L}$.

E

Therefore we may write $\dfrac{d^2v}{dx^2} = \dfrac{\omega}{2}(L-x)^2 \times \dfrac{L}{EI(3L-x)}$,

which reduces to $\dfrac{2EI}{\omega L}\dfrac{d^2v}{dx^2} = \dfrac{(L-x)^2}{(3L-x)}$.

In preparation for integration we put $\dfrac{(L-x)^2}{(3L-x)} = \dfrac{4L^2}{(3L-x)} - L - x$.

Integrating:

$$\frac{2EI}{\omega L}\frac{dv}{dx} = -4L^2 \log_e(3L-x) - Lx - \frac{x^2}{2} + C_1$$

$$\frac{2EI}{\omega L}v = 4L^2(3L-x)(\log_e(3L-x)-1) - \frac{Lx^2}{2} - \frac{x^3}{6} + C_1x + C_2.$$

Boundary conditions:

When $x = 0$, $\dfrac{dv}{dx} = 0$, therefore $C_1 = 4L^2 \log_e 3L$.

When $x = 0$, $v = 0$, therefore $C_2 = -12L^3(\log_e 3L - 1)$.

The displacement at the free end V_B can now be evaluated,

$$\frac{2EIV_B}{\omega L} = 8L^3(\log_e 2L - 1) - \frac{L^3}{2} - \frac{L^3}{6} + 4L^3 \log_e 3L - 12L^3(\log_e 3L - 1),$$

whence $V_B = \dfrac{\omega L^4}{EI}\left[\tfrac{5}{3} + 4\log_e \tfrac{2}{3}\right]\left(\text{approximately } 0.047\dfrac{\omega L^4}{EI}\right)$.

(a) *Discontinuity of loading*

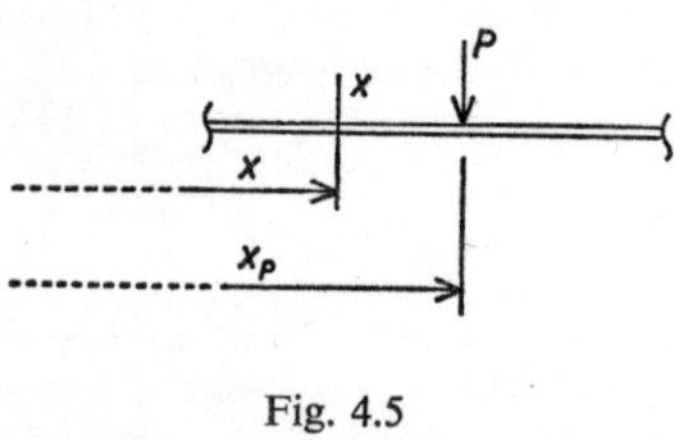

Fig. 4.5

Fig. 4.5 shows part of a flexural member in the region of a point load P at $x = x_p$ which will cause a discontinuity in the shear force and bending moment expressions at this point.

For $x \leqq x_p$ let the bending moment expression be $M = M_1$, a function of x.

For $x \geqq x_p$ the bending moment expressions will then be,

$$M = M_1 + P\langle x - x_p \rangle.$$

These two expressions may be integrated to give the slope and displacement functions either side of P as follows:

For $x \leqq x_p$,

$$EI\frac{dv}{dx} = \int M_1 dx + C_1,$$

$$EIv = \iint M_1 dx^2 + C_1 x + C_2.$$

For $x \geqq x_p$,

$$EI\frac{dv}{dx} = \int M_1 dx + \frac{P}{2}\langle x - x_p\rangle^2 + C_1'$$

$$EIv = \iint M_1 dx^2 + \frac{P}{6}\langle x - x_p\rangle^3 + C_1' x + C_2'.$$

At $x = x_p$ both pairs of equations must give the same values for $\frac{dv}{dx}$ and v. Hence we find $C_1 = C_1'$ and $C_2 = C_2'$. In other words the first pair of equations could have been obtained from the second, simply by ignoring the term involving P. This gives a convenient way of dealing with discontinuity of loading, in which the most general bending moment expression is integrated to give slope and displacement, non-relevant terms being ignored as appropriate when considering a particular position. It should be noted that the method relies on integrating the expressions such as $P\langle x - x_p\rangle$ as 'functions of functions' and not after expansion. That is,

$$\int P\langle x - x_p\rangle dx \rightarrow \frac{P}{2}\langle x - x_p\rangle^2 + C \textit{ not } \frac{Px^2}{2} - Px_p x + C.$$

The angle brackets indicate functions treated in this way. For discontinuities of bending moment, such as are produced by a couple, the same integration technique may be used. For a bending moment expression involving a couple T whose point of action is x_T the integration must be of the form:

$$\int T dx = T\langle x - x_T\rangle + C.$$

EXAMPLE 4.4

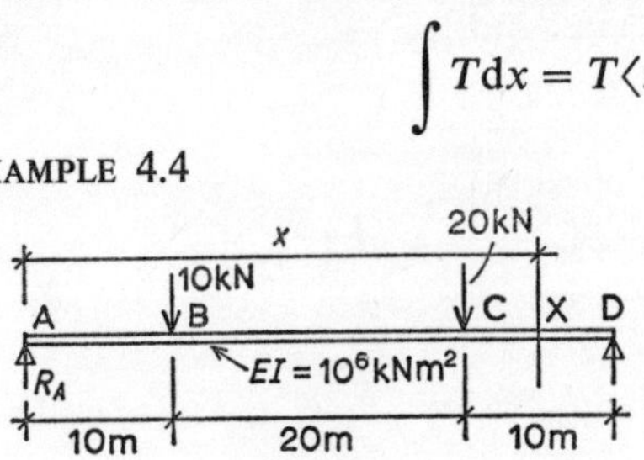

Fig. 4.6

$$R_A = \frac{20 \times 10 + 10 \times 30}{40} = 12.5 \text{ kN}.$$

The most general expression for the bending moment is obtained by considering the position X as shown.

$$M = -12.5x + 10\langle x - 10\rangle + 20\langle x - 30\rangle = 10^6 \frac{d^2 v}{dx^2}.$$

Integrating:

$$10^6 \frac{dv}{dx} = -\frac{12.5x^2}{2} + \frac{10}{2}\langle x-10\rangle^2 + \frac{20}{2}\langle x-30\rangle^2 + C_1,$$

$$10^6 v = -\frac{12.5x^3}{6} + \frac{10}{6}\langle x-10\rangle^3 + \frac{20}{6}\langle x-30\rangle^3 + C_1 x + C_2.$$

Boundary conditions:

When $x = 0$, $v = 0$, therefore $C_2 = 0$ ('point load' terms ignored).

When $x = 40$, $v = 0$, therefore

$$0 = -\frac{12.5}{6} \times 40^3 + \frac{10}{6} \times 30^3 + \frac{20}{6} \times 10^3 + 40C_1,$$

whence $C_1 = 2\,125$.

To find the maximum displacement we need to find the point at which $\frac{dv}{dx}$ is zero. In this example it is clearly between B and C, but in a more complicated case we might have to try several regions. If $\frac{dv}{dx}$ is zero between B and C we may write,

$$0 = -\frac{12.5}{2} x^2 + \frac{10}{2}\langle x-10\rangle^2 - 2\,125,$$

(the term involving the load at C is not required) the practical solution of which is $x = 20.8$, giving a maximum displacement value of $v = 0.027\,5$ m.

EXAMPLE 4.5

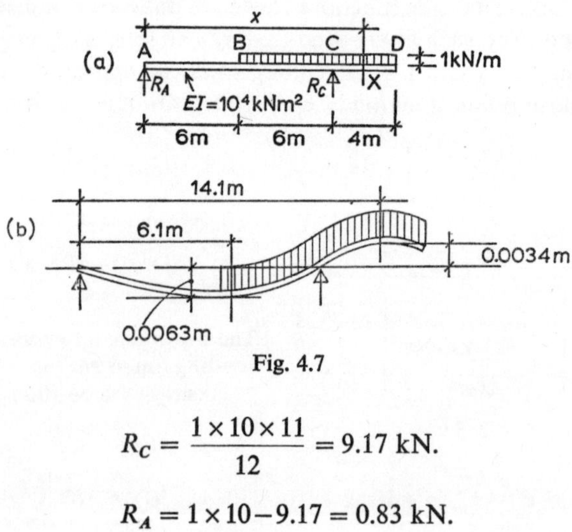

Fig. 4.7

$$R_C = \frac{1 \times 10 \times 11}{12} = 9.17 \text{ kN}.$$

$$R_A = 1 \times 10 - 9.17 = 0.83 \text{ kN}.$$

For X in the position shown,

$$M = -0.83x + \frac{\langle x-6\rangle^2}{2} - 9.17\langle x-12\rangle,$$

$$= 10^4 \frac{d^2v}{dx^2}.$$

Integrating:

$$10^4 \frac{dv}{dx} = -\frac{0.83x^2}{2} + \frac{+\langle x-6\rangle^3}{6} - \frac{9.17\langle x-12\rangle^2}{2} + C_1,$$

$$10^4 v = -\frac{0.83x^3}{6} + \frac{\langle x-6\rangle^4}{24} - \frac{9.17\langle x-12\rangle^3}{6} + C_1x + C_2.$$

Boundary conditions:

When $x = 0$, $v = 0$, therefore $C_2 = 0$ (distributed load and R_c terms ignored).

When $x = 12$, $v = 0$, therefore $0 = -\frac{0.83 \times 12^3}{6} + \frac{6^4}{24} + 12C_1$,

whence $C_1 = 15.4$.

It is not immediately apparent where the maximum displacement occurs, but there will obviously be a stationary value either between A and B, or between B and C.

For a stationary value between A and B,

$$0 = -\frac{0.83x^2}{2} + 15.4, \text{ whence } x = 6.09 \text{ m}.$$

This value of x is outside the region A-B and so we conclude that there is no stationary value in this region.

For a stationary value between B and C,

$$0 = -\frac{0.83x^2}{2} + \frac{\langle x-6\rangle^3}{6} + 15.4, \text{ whence } x = 6.1$$

(solving by trial and error, or graphically for feasible answers).

At this point the displacement is given by

$$v = \frac{1}{10^4}\left(-\frac{0.83 \times 6.1^3}{6} + \frac{0.1^4}{24} + 15.4 \times 6.1\right) = 0.006\,3.$$

There may be a stationary value between C and D, in which case,

$$0 = -\frac{0.83x^2}{2} + \frac{\langle x-6\rangle^3}{6} - \frac{9.17\langle x-12\rangle^2}{2} + 15.4,$$

whence $x = 14.1$.

At this point the displacement is given by,

$$v = \frac{1}{10^4}\left(\frac{-0.83 \times 14.1^3}{6} + \frac{8.1^4}{24} - \frac{9.17 \times 2.1^3}{2} + 15.4 \times 14.1\right)$$

$$= -0.003\,4.$$

The displaced shape of the member as shown in Fig. 4.7 (b).

(b) *Discontinuity of member properties*

Discontinuities such as a change of section or the presence of a hinge cannot be treated in the same way as discontinuities of loading. Separate slope and displacement equations may be found for each section, which must be compatible at the point of discontinuity.

EXAMPLE 4.6

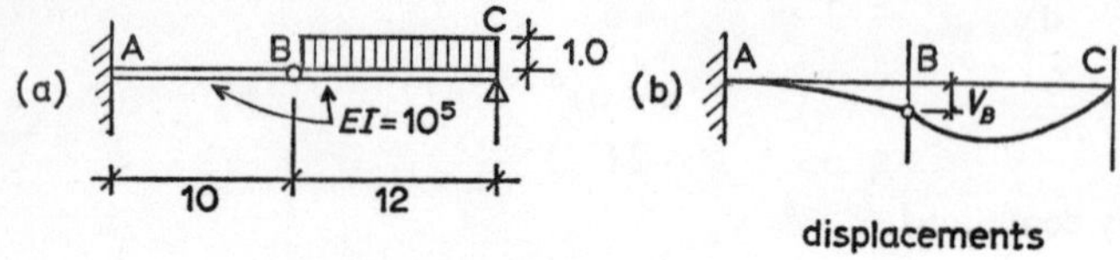

Fig. 4.8 (a) and (b)

Here the discontinuity of slope at B makes it impossible to arrive at the equation of slope and displacement by direct integration from a single bending moment expression. The value of V_B may be found by considering the displacements of A-B separately, and this may then be used as a boundary displacement when considering B-C.

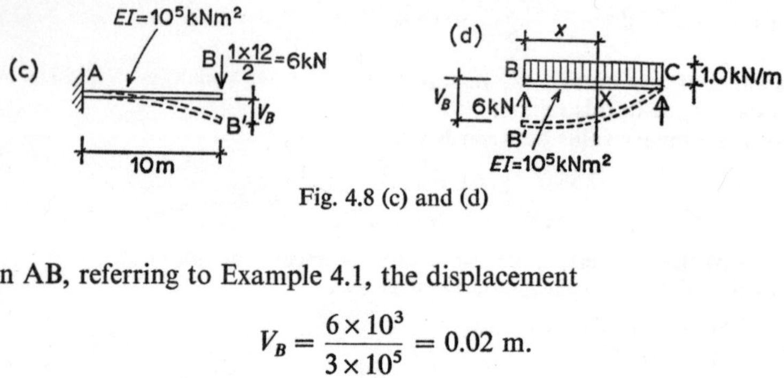

Fig. 4.8 (c) and (d)

In AB, referring to Example 4.1, the displacement

$$V_B = \frac{6 \times 10^3}{3 \times 10^5} = 0.02 \text{ m.}$$

For a general position X in BC we have.

$$10^5 \frac{d^2v}{dx^2} = \frac{x^2}{2} - 6x.$$

Integrating:

$$10^5 \frac{dv}{dx} = \frac{x^3}{6} - \frac{6x^2}{2} + C_1,$$

$$10^5 v = \frac{x^4}{24} - \frac{6x^3}{6} + C_1 x + C_2.$$

Boundary conditions:

When $x = 0$, $v = 0.02$, therefore $C_2 = 2\,000$.

When $x = 12$, $v = 0$, therefore $0 = 864 - 1\,728 + C_1 \times 12 + 2\,000$, whence $C_1 = -94.7$.

A stationary value of v will occur when $\dfrac{dv}{dx} = 0$, that is when

$$0 = \frac{x^3}{6} - 3x^2 - 94.7.$$

There is no value of x between 0 and 12 which satisfies this equation and so we conclude that the section B-C does not sag below the level of B′ as indicated in (b) and (d). The maximum displacement from the straight line B′C may be found using the result of Example 4.2, from which $v'_{max} = 0.0027$ m.

EXAMPLE 4.7

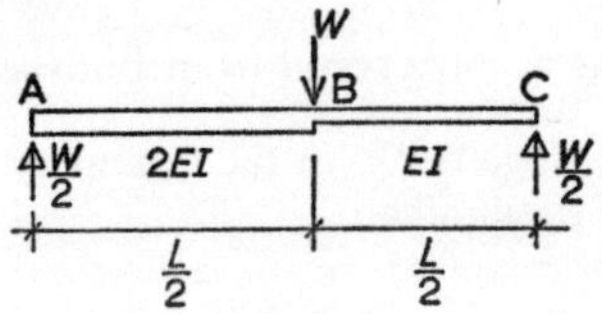

Fig. 4.9

For a general position in the region A-B, at x from A, the bending moment is $-\dfrac{W}{2}x$.

Thus,
$$2EI\frac{d^2v}{dx^2} = -\frac{W}{2}x.$$

Integrating:
$$2EI\frac{dv}{dx} = -\frac{Wx^2}{4} + C_1$$

$$2EIv = -\frac{Wx^3}{12} + C_1x + C_2.$$

Boundary conditions:

at $x = 0$, $v = 0$, therefore $C_2 = 0$

at B, we have
$$\left(\frac{dv}{dx}\right)_B = \frac{1}{2EI}\left(-\frac{WL^2}{16} + C_1\right)$$

and
$$v_B = \frac{1}{2EI}\left(-\frac{WL^3}{96} + \frac{C_1L}{2}\right).$$

For a general position in the region B-C, at x from B, we have

$$EI\frac{d^2v}{dx^2} = Wx - \frac{W}{2}\left(x + \frac{L}{2}\right).$$

Integrating: $$EI\frac{dv}{dx} = \frac{Wx^2}{2} - \frac{W}{4}\left(x+\frac{L}{2}\right)^2 + C_3$$

$$EIv = \frac{Wx^3}{6} - \frac{W}{12}\left(x+\frac{L}{2}\right)^3 + C_3x + C_4.$$

Boundary conditions:

at $x = \frac{L}{2}$, $v = 0$, therefore $\frac{C_3L}{2} + C_4 = \frac{WL^3}{16}$

at B, we have $$\left(\frac{dv}{dx}\right)_B = \frac{1}{EI}\left(-\frac{WL^2}{16} + C_3\right)$$

and $$v_B = \frac{1}{EI}\left(-\frac{WL^3}{96} + C_4\right).$$

Clearly the same values for $\left(\frac{dv}{dx}\right)_B$ and v_B must result from both sets of equations and these *compatibility* conditions, together with the boundary conditions at A and C enable C_1, C_2, C_3, C_4 to be determined.

The values are found to be as follows:

$$C_1 = \frac{WL^2}{12}, \quad C_2 = 0$$

$$C_3 = \frac{7WL^2}{96}, \quad C_4 = \frac{5WL^3}{192}.$$

The slope and displacement at B are found to be:

$$\left(\frac{dv}{dx}\right)_B = \frac{WL^2}{96EI}$$

$$v_B = \frac{WL^3}{64EI}.$$

(c) *Non-rectilinear members*

It is not usually practicable to use this 'direct' approach to determine the displacements of curved members; these are usually obtained by 'energy' methods,* which are outside the scope of this work. Members which consist of a series of straight pieces, however, can sometimes be dealt with quite simply, as in the following example.

* *Virtual Work and Energy Concepts*, J. Rhodes (Chatto and Windus) 1974.

EXAMPLE 4.8

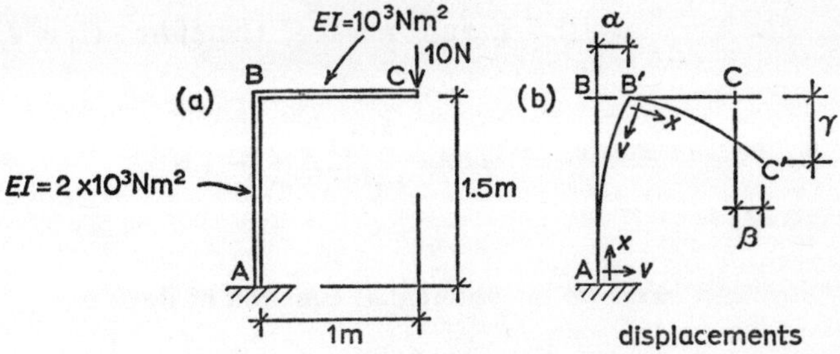

Fig. 4.10

Since we are considering only 'small' displacements it follows that the displacement BB′ is horizontal to the 1st order of approximation and that $\alpha = \beta$. The corner B remains a right angle during displacement, and so the slope of AB′ at B′ is the same as that of B′C′ at that point.

For the part AB, measuring x from A we have

$$2\times 10^3 \frac{d^2v}{dx^2} = 10\times 1.$$

Integrating:
$$2\times 10^2 \frac{dv}{dx} = x + C_1$$

$$2\times 10^2 v = \frac{x^2}{2} + C_1x + C_2.$$

Boundary conditions:

When $x = 0$, $\frac{dv}{dx} = 0$ and $v = 0$; therefore $C_1 = C_2 = 0$.

At B', then, $\frac{dv}{dx} = 0{\cdot}007\,5$ and $v = 0{\cdot}005\,63$ $(= \alpha)$.

For the part BC, measuring x from B we have

$$10^3 \frac{d^2v}{dx^2} = 10(1-x).$$

Integrating:
$$10^3 \frac{dv}{dx} = \frac{10}{2}(1-x)^2 + C_3$$

$$10^3 v = \frac{10}{6}(1-x)^3 + C_3x + C_4.$$

Boundary conditions:

When $x = 0$, $\frac{dv}{dx} = 0.007\,5$ and $v = 0$. Therefore $C_3 = 2.5$ and $C_4 = -1.67$.

The vertical displacement at C is thus $\frac{1}{10^3}(2.5 \times 1 - 1.67)$, that is 0.000 83 ($= \gamma$).

4.2 Two theorems based on the differential equation of flexure

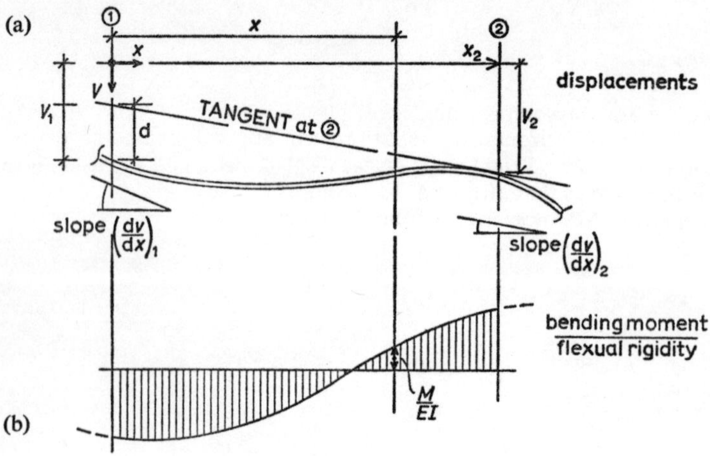

Fig. 4.11

For the general case when both M and EI vary, the differential equation of flexure may be written

$$\frac{d^2v}{dx^2} = \left(\frac{M}{EI}\right), \text{ a function of } x.$$

Integrating both sides with respect to x over the region ①-② we have,

$$\left[\frac{dv}{dx}\right]_{①}^{②} = \int_{①}^{②} \left(\frac{M}{EI}\right) dx,$$

that is,

$$\left(\frac{dv}{dx}\right)_2 - \left(\frac{dv}{dx}\right)_1 = \int_{①}^{②} \left(\frac{M}{EI}\right) dx. \tag{4.5}$$

In other words,

The change of slope between two points is given by the area of the $\frac{M}{EI}$ *diagram between them.*

If both sides of the equation of flexure are multiplied by x before integrating we obtain the following result,

$$\int_{①}^{②}\left(\frac{d^2v}{dx^2}\right)x\,dx = \int_{①}^{②}\left(\frac{M}{EI}\right)x\,dx.$$

Performing the left-hand integral by parts we have

$$\left[\left(\frac{dv}{dx}\right)x - \int\frac{dv}{dx}\right]_{①}^{②} = \int_{①}^{②}\left(\frac{M}{EI}\right)x\,dx$$

which leads to

$$\left(\frac{dv}{dx}\right)_2 x_2 - v_2 + v_1 = \int_{①}^{②}\left(\frac{M}{EI}\right)x\,dx. \qquad (4.6)$$

Referring to the diagrams it can be seen that the left-hand side of equation (4.6) is in fact the dimension d; the right-hand side represents the moment of the $\frac{M}{EI}$ diagram about ①. In other words,

The displacement of one point measured from a tangent drawn at a second point is given by the moment of the $\frac{M}{EI}$ *diagram between these points, about the first point.*

These two theorems, often referred to as the *area moment theorems* (attributed to Mohr) provide a very neat way of solving certain beam flexure problems, when slopes and displacements at only a few points are required. They are also particularly useful when tackling some of the statically indeterminate problems dealt with in § 4.3.

EXAMPLE 4.9

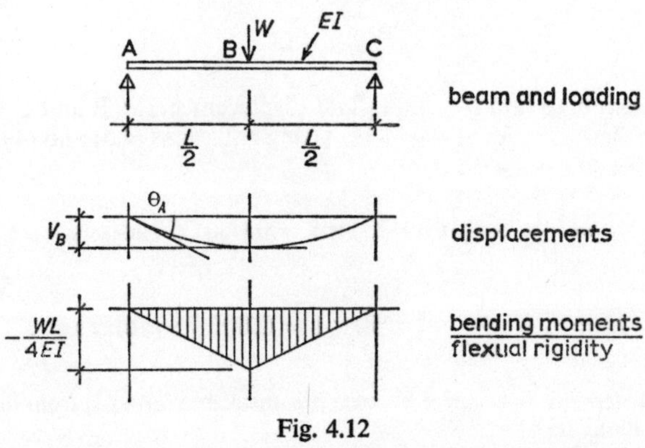

Fig. 4.12

To find the slope at A we note that by symmetry the slope at B is zero. Then, applying the first theorem we have

$$0-\theta_A = \tfrac{1}{2}\times\left(-\frac{WL}{4EI}\right)\times\frac{L}{2},$$

whence
$$\theta_A = \frac{WL^2}{16EI}.$$

To find the displacement at B we note that this is equal to the (negative)* displacement at A from a tangent to the beam at B. Applying the second theorem we obtain

$$-v_B = \tfrac{1}{2}\times\left(-\frac{WL}{4EI}\right)\times\frac{L}{2}\times\frac{L}{3},$$

that is,
$$v_B = \frac{WL^3}{48EI}.$$

EXAMPLE 4.10

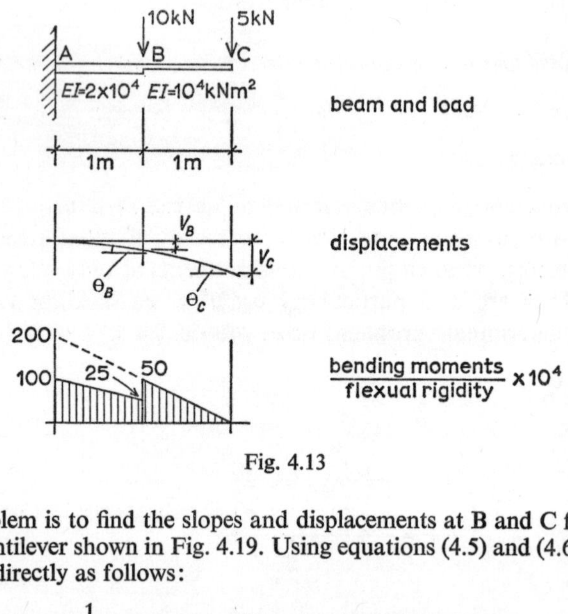

Fig. 4.13

The problem is to find the slopes and displacements at B and C for the non-uniform cantilever shown in Fig. 4.19. Using equations (4.5) and (4.6) these may be written directly as follows:

$$\theta_B = \frac{1}{10^4}\times\tfrac{1}{2}(100+25)\times 1 = 0.006\,25 \text{ radians}$$

$$\theta_C = \theta_B + \frac{1}{10^4}\times\tfrac{1}{2}\times 50\times 1 = 0.008\,75 \text{ radians}$$

* This displacement is negative because it is measured upwards from the tangent, in the opposite direction to v.

$$v_B = \frac{1}{10^4} \times \tfrac{1}{6}(2\times 100+25)\times 1^2 = 0.003\ 75\ \text{m}$$

$$v_C = v_B + 1\times\theta_B + \frac{1}{10^4} \times \tfrac{1}{3}\times 50\times 1^2 = 0.011\ 67\ \text{m}.$$

PROBLEMS

(answers given at the end of the chapter)

PROBLEMS 4.1-4.8

Sketch the displaced form of the following members, calculating values at sufficient points to enable you to do so (including stationary values). Units are kilonewtons and metres.

PROBLEM 4.1

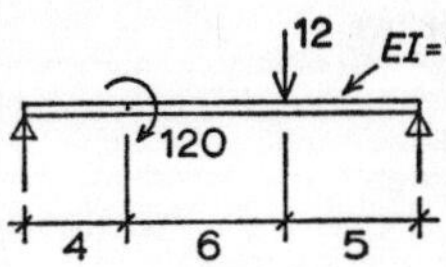

PROBLEM 4.2

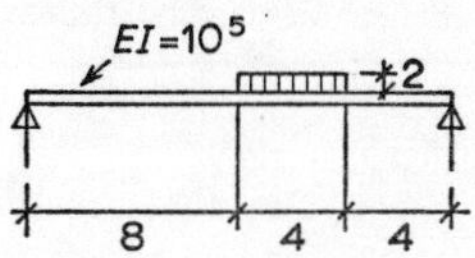

PROBLEM 4.3

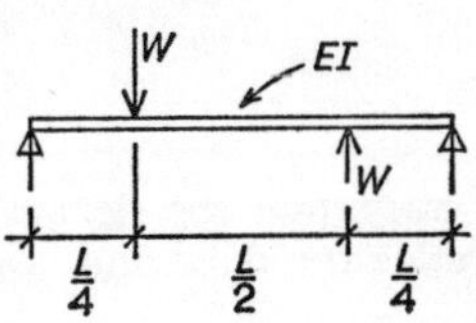

PROBLEM 4.4

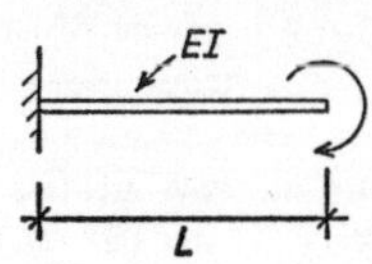

PROBLEM 4.5

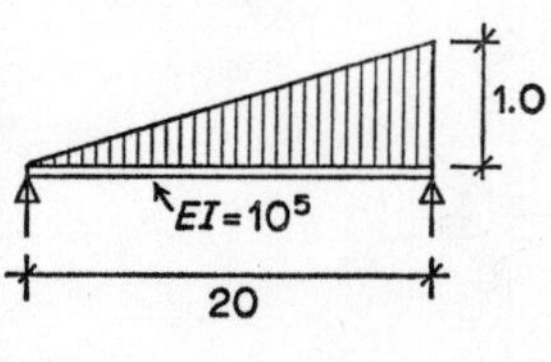

PROBLEM 4.6

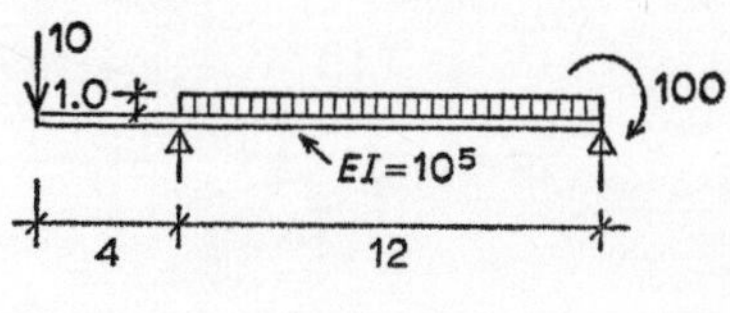

PROBLEM 4.7

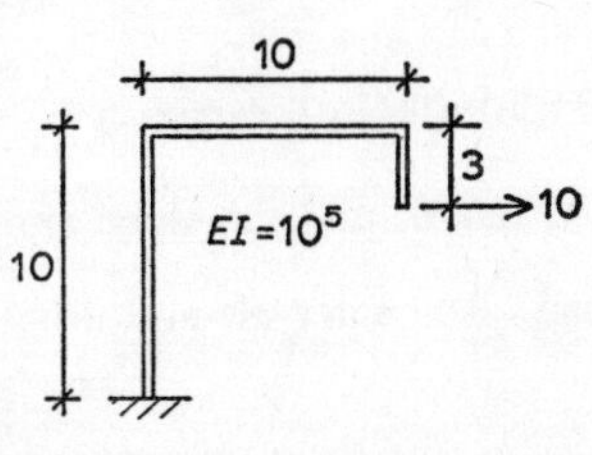

PROBLEM 4.8

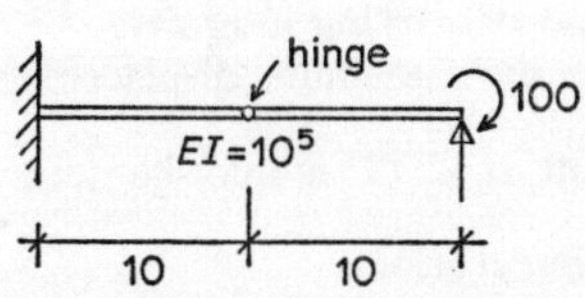

PROBLEM 4.9

By considering the effect of a single point load, show that the maximum displacement of a simple uniform beam of span L, subjected to vertical loads only (point or distributed), occurs never more than 0.077 L from mid-span.

PROBLEM 4.10

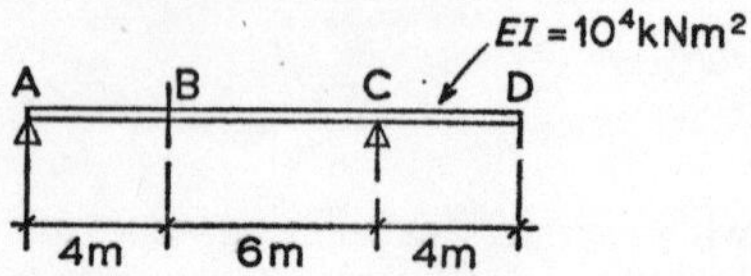

Plot the influence line for the displacement at B.

PROBLEM 4.11

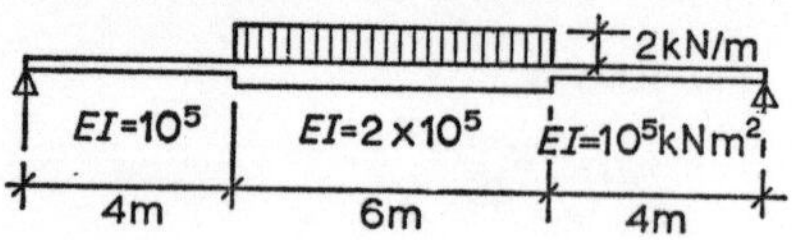

Determine the maximum displacement of this beam.

PROBLEM 4.12

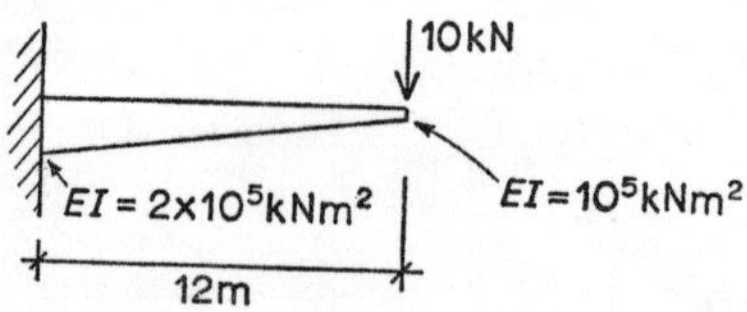

The cantilever beam shown in the diagram tapers such that the flexural rigidity varies linearly from 2×10^5 to 10^5 kNm^2. Determine the slope and displacement at the free end.

PROBLEM 4.13

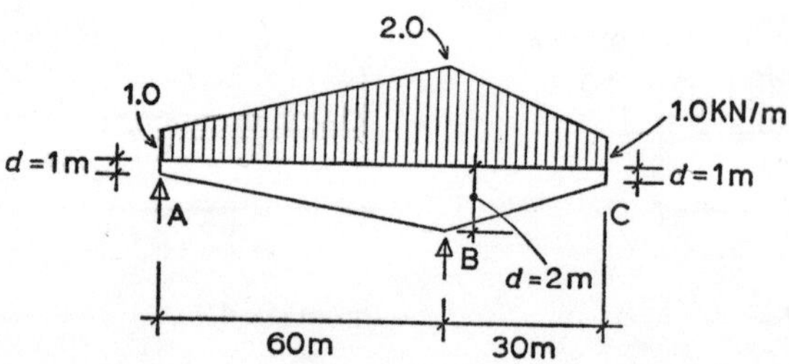

The non-uniform beam above, for which EI$=4 \times 10^6 d^3$kNm^2, carries a distributed load of varying intensity.

Calculate the maximum displacement between A and B, and the slope and displacement at C. (This may be done by evaluating $\dfrac{M}{EI}$ at intervals and using numerical integration.)

PROBLEM 4.14

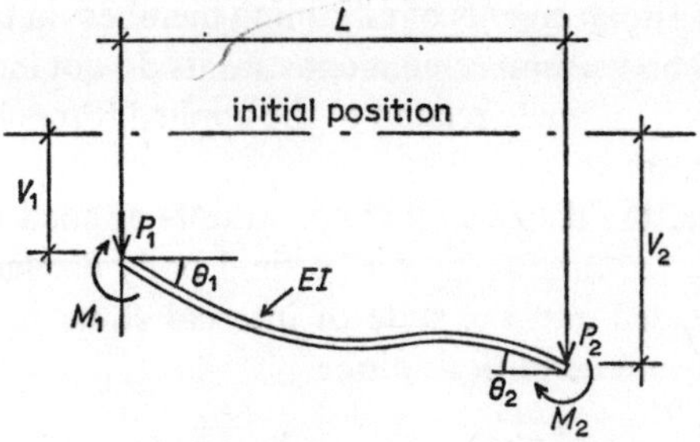

Establish the relationships between the forces, moments, displacements and rotations for the member shown in the figure, to obtain the coefficients of the 'stiffness matrix' in the following equations:

$$\begin{bmatrix} P_1 \\ M_1 \\ P_2 \\ M_2 \end{bmatrix} = \begin{bmatrix} 4\times 4 \\ \text{'stiffness} \\ \text{matrix'} \end{bmatrix} \times \begin{bmatrix} v_1 \\ \theta_1 \\ v_2 \\ \theta_2 \end{bmatrix}.$$

4.3 The solution of statically indeterminate problems

The general displacements of a completely unconstrained member may be regarded as composed of two quite distinct components, *rigid body displacements* and *deformations*. Rigid body displacements are the translation and rotation which would result if the member were completely inflexible. The deformations are those displacement components which represent a change of shape and are uniquely determined by the state of internal stress.

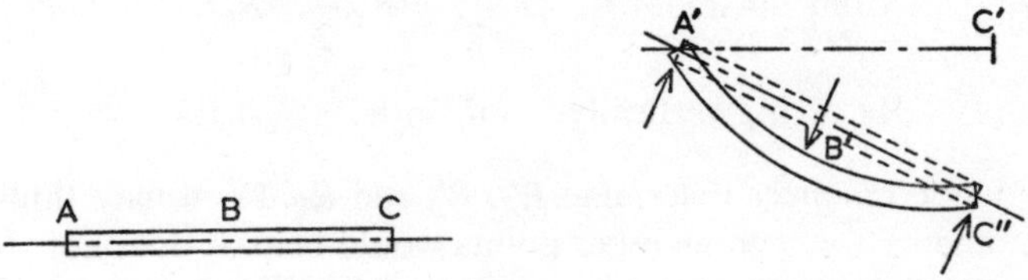

Fig. 4.14

In Fig. 4.14 the *rigid body displacement* consists of the translation of AC to A′C′ and the rotation of A′C′ to A′C″ of the original shape of the member as indicated by the broken line. The *deformation* is the displacement from the original straight line A′C″ to the curve A′B′C″.

In most structural applications we are not concerned with members possessing complete freedom of displacement, rigid body displacements being either completely prevented as in the case of isolated members, or prescribed by the displacements of adjoining members in the case of frameworks. Such rigid body displacement constraints do not induce any internal stresses, these being uniquely related to the applied forces by the conditions of static equilibrium.

When, however, the displacement constraints applied to a member are in excess of those required to prescribe rigid body displacements, stresses are thereby generated and the state of internal stress is no longer determined by equilibrium conditions alone.

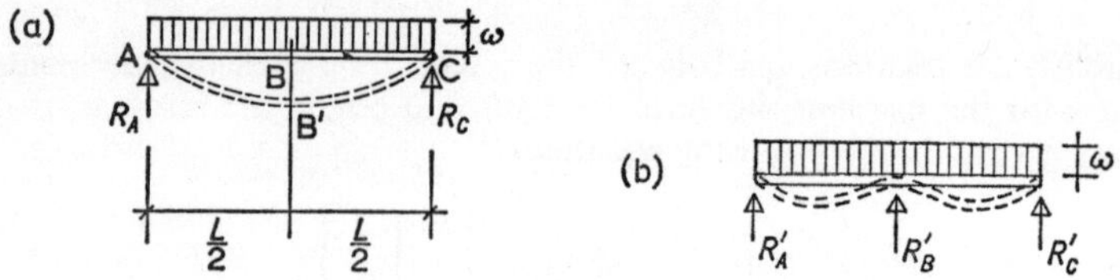

Fig. 4.15

In Fig. 4.15 (a) rigid body displacements of the beam ABC are prevented by prescribing zero displacement at A and C so that the deformed shape of the beam is AB′C as shown.

If in addition we prescribe zero displacement at B the deformed shape is as shown in (b) and clearly the state of internal stress is quite different.

In case (a) the conditions of statical equilibrium alone determine the reactions R_A and R_C, each being $\frac{\omega L}{2}$, and hence the stress resultants. In case (b) the (two) conditions of statical equilibrium may be written

$$\text{Moments about A,}\quad \frac{\omega L^2}{2} = R'_B\frac{L}{2} + R'_C L$$

$$\text{Resolving vertically,}\quad \omega L = R'_A + R'_B + R'_C$$

which do not uniquely determine R'_A, R'_B and R'_C. (You may think that taking moments about some other points would help. It does not; but do try it, to convince yourself.)

Only by considering the deformation effects of the additional constraint at B can the problem be solved.

Problems of this type are known as *statically indeterminate*, the number of constraints in excess of those necessary to prescribe rigid body displacements being called the *degree of indeterminacy*. In the simple example illustrated in Fig. 4.15 the one 'excess' constraint is *external* and is associated with the reaction R_B. However, the same final result could have been arrived at in a slightly different way, involving an *internal* constraint as shown in Fig. 4.16.

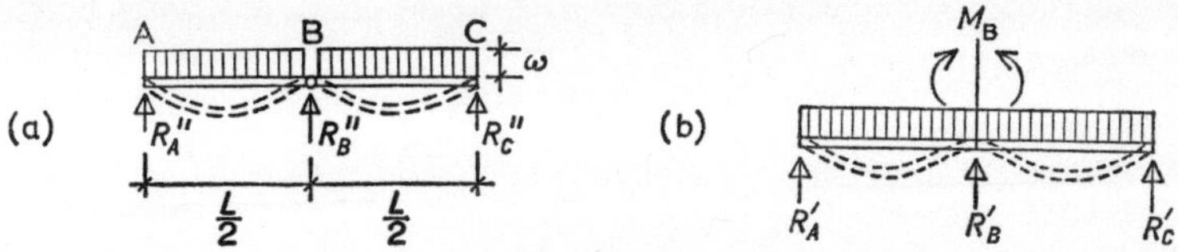

Fig. 4.16

The hinged member ABC in (a) is statically determinate, the reactions being simply found by considering the equilibrium of the whole member and one of the parts. It is clear that there is relative rotation of the parts of the member meeting at the pin joint, each span of the beam sagging independently under the load. In (b) an *internal* displacement constraint is introduced, by which the relative rotation is prevented and this is clearly associated with the bending moment M_B. We now have precisely the same situation as in 4.15 (b).

One way of solving statically indeterminate problems is to treat the forces or moments associated with the 'excess' displacement constraints as the unknowns and to express the internal stresses in terms of these unknowns. Deformations may then be determined, again in terms of the unknowns, which may then be found by applying the 'excess' displacement constraints, or *compatibility conditions*.

The following examples illustrate the use of the equation of flexure and the theorems derived from it to solve statically indeterminate problems in this way.

EXAMPLE 4.11

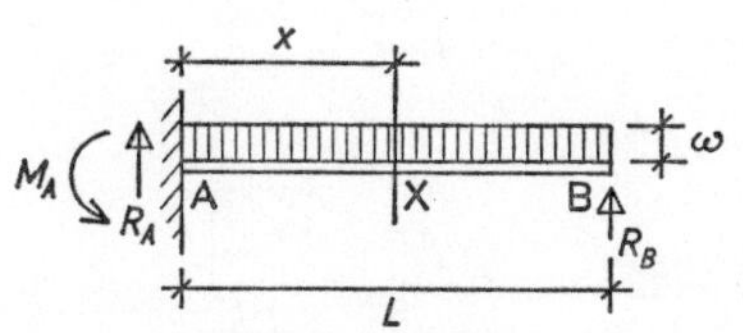

Fig. 4.17 (a)

The boundary conditions of the member are:

zero slope at A corresponding to reaction M_A
zero displacement at A corresponding to reaction R_A
zero displacement at B corresponding to reaction R_B.

The number of *distinct* equilibrium equations which can be written for the member AB (in the absence of axial forces and displacements) is *two*, and so the number of boundary conditions is in excess of that required for statical equilibrium. This means that M_A, R_A and R_B cannot be determined merely by taking moments and resolving; the most that can be achieved is to express any two in terms of the third. However, if we write the shear force or bending moment expressions in terms of one of these unknowns, and then proceed by integration to obtain the slope and displacement equations, the 'extra' boundary condition will enable this unknown reaction to be determined. Choosing R_B as the unknown,

at X, the bending moment $= \frac{\omega}{2}(L-x)^2 - R_B(L-x) = EI\frac{d^2v}{dx^2}$.

Integrating: $$EI\frac{dv}{dx} = \frac{\omega}{6}(L-x)^3 - \frac{R_B}{2}(L-x)^2 + C_1,$$

$$EIv = \frac{\omega}{24}(L-x)^4 - \frac{R_B}{6}(L-x)^3 + C_1x + C_2.$$

Boundary conditions:

at A, $x = 0$, $\frac{dv}{dx} = 0$, therefore $C_1 = \frac{R_B}{2}L^2 - \frac{\omega L^3}{6}$,

also $v = 0$, therefore $C_2 = \frac{R_B}{6}L^3 - \frac{\omega L^4}{24}$,

at B, $x = L$, $v = 0$, therefore $C_1L + C_2 = 0$.

Solving these equations gives $$R_B = \frac{3}{8}\omega L,$$

$$C_1 = -\frac{\omega L^3}{48},$$

$$C_2 = \frac{\omega L^4}{48}.$$

Considering the equilibrium of the member will enable the remaining two reactions to be found,

$$M_A = \frac{\omega L^2}{2} - R_B L = \frac{\omega L^2}{8},$$

$$R_A = \omega L - R_B = \frac{5\omega L}{8}.$$

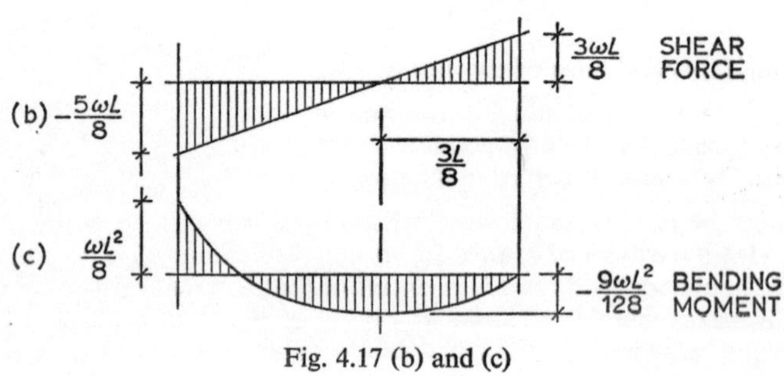

Fig. 4.17 (b) and (c)

The displacement equation is,

$$EIv = \frac{\omega}{24}(L-x)^4 - \frac{3\omega L}{48}(L-x)^3 - \frac{\omega L^3 x}{48} + \frac{\omega L^4}{48}.$$

There is a stationary point when,

$$\frac{dv}{dx} = \frac{\omega}{24}(L-x)^3 - \frac{3\omega L}{16}(L-x)^2 - \frac{\omega L^3}{48} = 0,$$

from which we obtain a value of 0.654 L for x.

At this position $$v_{max} = 0.011\,6\,\frac{\omega L^4}{EI}.$$

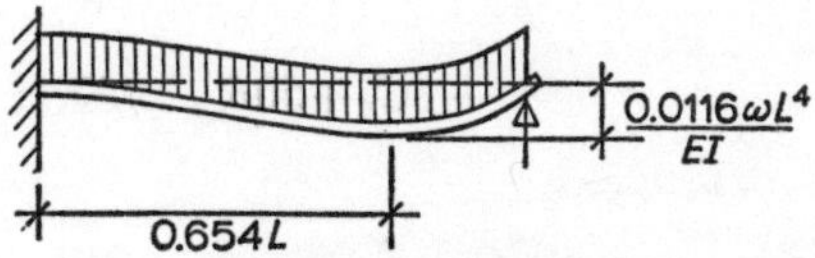

DISPLACEMENTS
Fig. 4.17 (d)

The stress resultants have been plotted in Fig. 4.17 (b) and (c), and displacements in (d).

EXAMPLE 4.12

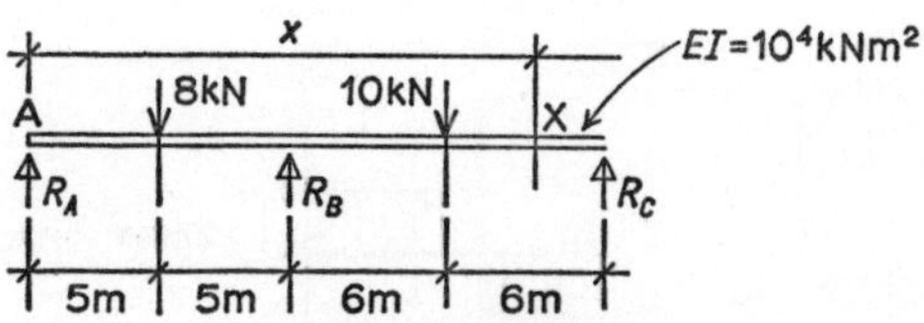

Fig. 4.18 (a)

In this problem there are three boundary conditions, corresponding to the reactions. R_A, R_B and R_C, which constitutes *one degree of indeterminacy*. R_A is chosen as the unknown, and it will be necessary to express R_B in terms of R_A for the purposes of obtaining a general expression for the bending moment at X.

$$R_B = \frac{1}{12}(10\times 6 + 8\times 17 - R_A\times 22)$$

$$= \frac{(98-11R_A)}{6}.$$

At X the bending moment is

$$-R_A x + 8(x-5) - \frac{(98-11R_A)}{6}(x-10) + 10(x-16) = 10^4\,\frac{d^2v}{dx^2}.$$

Integrating:

$$10^4\frac{dv}{dx} = -R_A\frac{x^2}{2}+\frac{8}{2}(x-5)^2-\frac{(98-11R_A)}{12}(x-10)^2+\frac{10}{2}(x-16)^2+C_1,$$

$$10^4 v = -R_A\frac{x^3}{6}+\frac{8}{6}(x-5)^3-\frac{(98-11R_A)}{36}(x-10)^3+\frac{10}{6}(x-16)^3 +C_1x+C_2.$$

Boundary conditions:

at A, $x = 0$, $v = 0$, therefore $C_2 = 0$,

at B, $x = 10$, $v = 0$, therefore $-\dfrac{R_A\times 10^3}{6}+\dfrac{8\times 5^3}{6}+C_1\times 10 = 0,$

whence $$-3C_1+50R_A = 50,$$

at C, $x = 22$, $v = 0$, therefore

$$-\frac{R_A\times 22^3}{6}+\frac{8\times 17^3}{6}-\frac{(98-11R_A)}{36}\times 12^3+\frac{10}{6}\times 6^3+C_1\times 22 = 0,$$

whence $$66C_1-3\,740R_A = -6\,620.$$

Solving the latter two equations gives, $C_1 = 18.2$ and $R_A = 2.09$.

It follows that $R_B = 12.50$ and $R_C = 3.41$ kN.

The stress resultant diagrams are as follows:

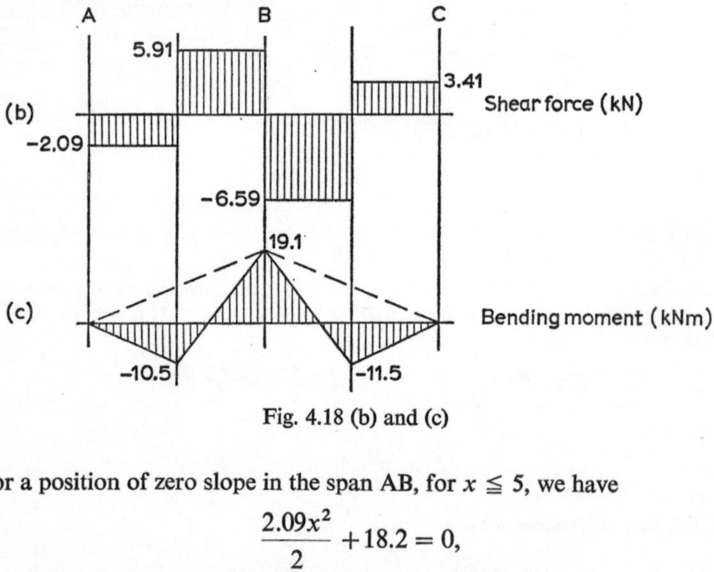

Fig. 4.18 (b) and (c)

For a position of zero slope in the span AB, for $x \leqq 5$, we have

$$\frac{2.09x^2}{2}+18.2 = 0,$$

for which a solution is $x = 4.17$ m, giving $v = 50.7\times 10^{-4}$ m.

For $x > 5$ we have $$-\frac{2.09x^2}{2}+4(x-5)^2+18.2=0,$$

for which a solution is $x = 9.21$ m, giving $v = -4.0\times10^{-4}$ m.

For a position of zero slope in the span BC, for $x \geqq 16$, we have

$$-\frac{2.09x^2}{2}+4(x-5)^2-\frac{12.5}{2}(x-10)^2+5(x-16)^2+18.2=0,$$

for which a solution is $x = 16.44$ m giving $v = 196\times10^{-4}$ m.

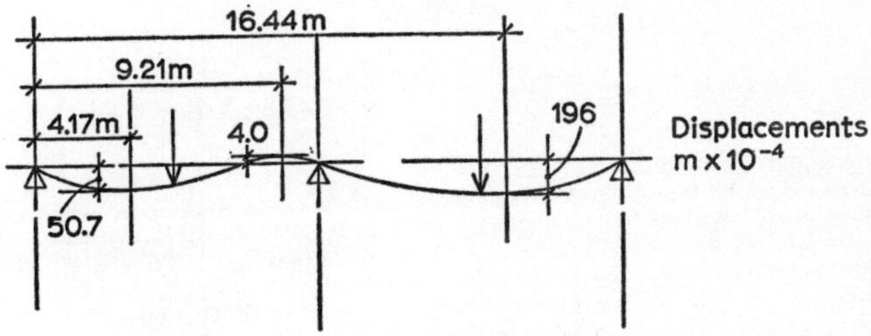

Fig. 4.18 (d)

The stress resultants have been plotted in Fig. 4.18 (b) and (c), and displacements in (d)

EXAMPLE 4.13

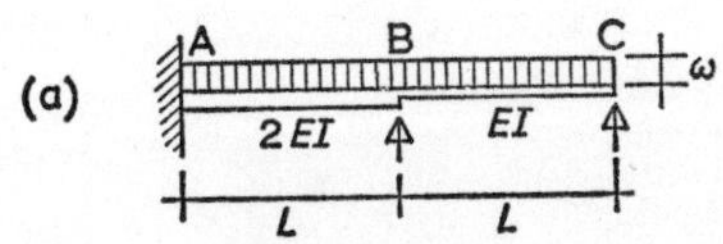

Fig. 4.19 (a)

Here there are *four* boundary conditions corresponding to *two* 'excess' constraints or degrees of indeterminacy. The problem will be solved by choosing the bending moments at A and B as the unknowns and using the 'area moment' theorems derived in § 4.2.

For convenience the effects of the applied loading and of the bending moments M_A and M_B are considered separately, in (b), (c) and (d).

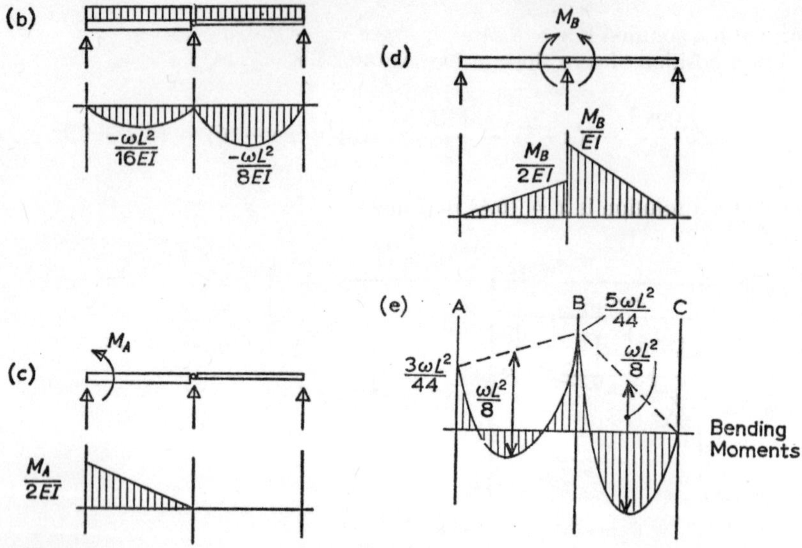

Fig. 4.19 (b), (c), (d) and (e)

Because the beam is encastré at A it follows that a tangent to the beam at this point would pass through both B and C. In other words, the displacements of B and C from the tangent at A are zero. This means that the corresponding moments of the $\frac{M}{EI}$ diagrams are also zero.

For zero moment about B we have

$$-\frac{2}{3}\frac{\omega L^3}{16EI}\times\frac{L}{2}+\frac{1}{2}\frac{M_A L}{2EI}\times\frac{2L}{3}+\frac{1}{2}\frac{M_B L}{2EI}\times\frac{L}{3}=0,$$

and for zero moment about C,

$$-\frac{2}{3}\frac{\omega L^3}{16EI}\times\frac{3L}{2}-\frac{2}{3}\frac{\omega L^3}{8EI}\times\frac{L}{2}+\frac{1}{2}\frac{M_A L}{2EI}\times\frac{5L}{3}$$
$$+\frac{1}{2}\frac{M_B L}{2EI}\times\frac{4L}{3}+\frac{1}{2}\frac{M_B L}{EI}\times\frac{2L}{3}=0.$$

Solving these two equations gives

$$M_A=\frac{3\omega L^2}{44} \text{ and } M_B=\frac{5\omega L^2}{44},$$

and the resulting (combined) bending moment diagram is given in (e).

The use of this method clearly enables the stress resultant distribution to be determined very quickly, but nothing is found about the displacements.

Problems

Problem 4.15

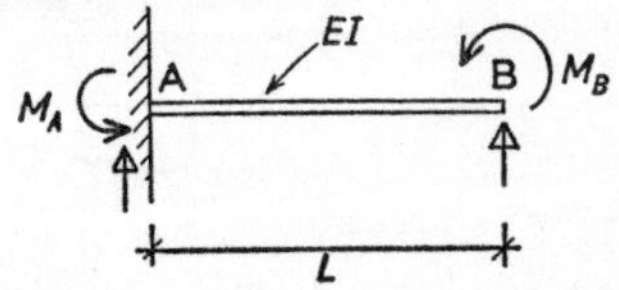

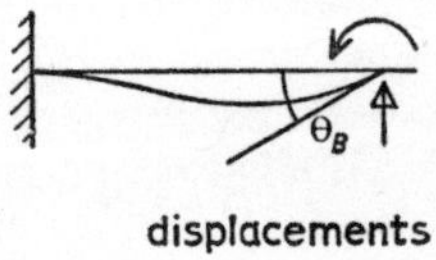

Prove that $M_A = \dfrac{M_B}{2}$, and $\theta_B = \dfrac{M_B L}{4EI}$.

Problem 4.16

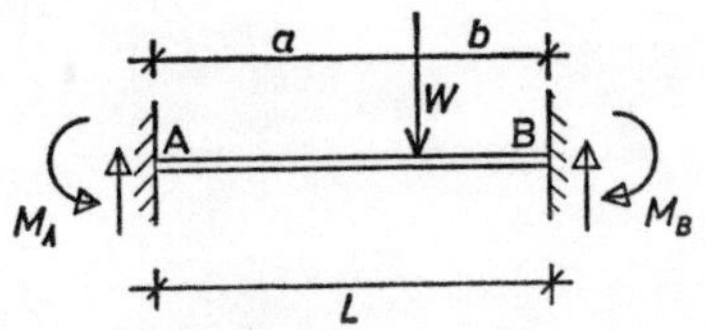

Prove that $M_A = \dfrac{Wab^2}{L^2}$

and $M_B = \dfrac{Wba^2}{L^2}$.

Problem 4.17

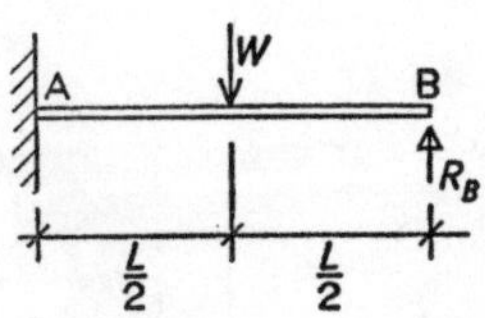

Prove that $R_B = \dfrac{5W}{16}$.

Problem 4.18

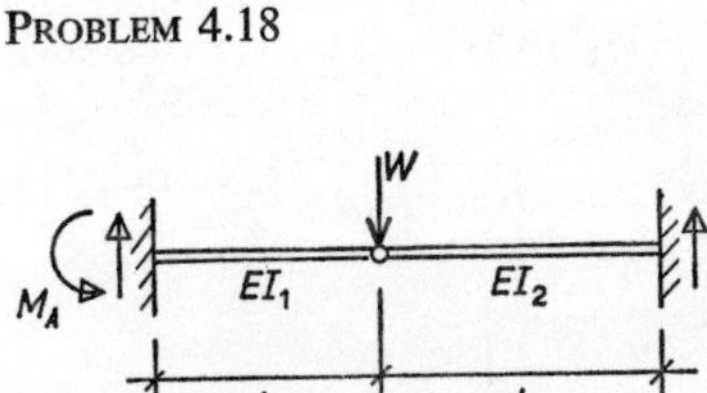

Prove that $M_A = \dfrac{Wk_1L_1}{(k_1+k_2)}$

and $M_B = \dfrac{Wk_2L_2}{(k_1+k_2)}$,

where $k_1 = \dfrac{EI_1}{L_1^3}$ and $k_2 = \dfrac{EI_2}{L_2^3}$.

ANSWERS TO PROBLEMS 4.1-4.14

4.1

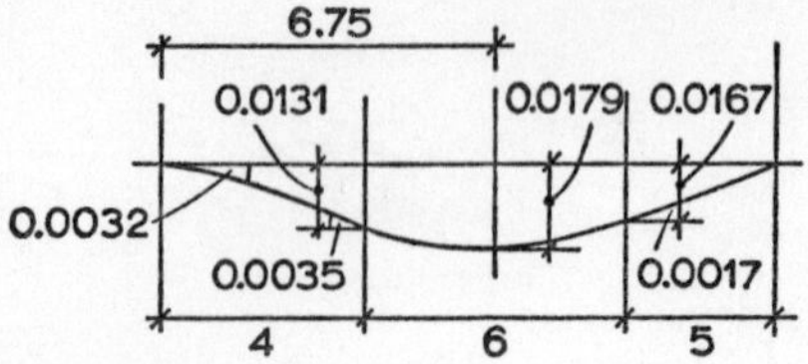

4.2

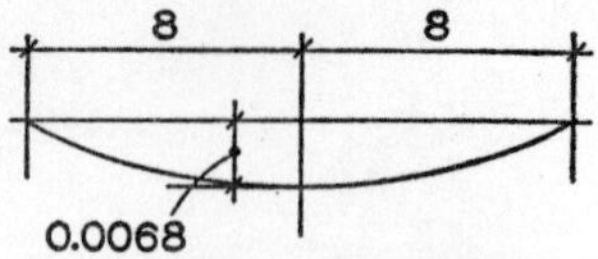

Note. maximum displacement occurs just to right of centre line

4.3

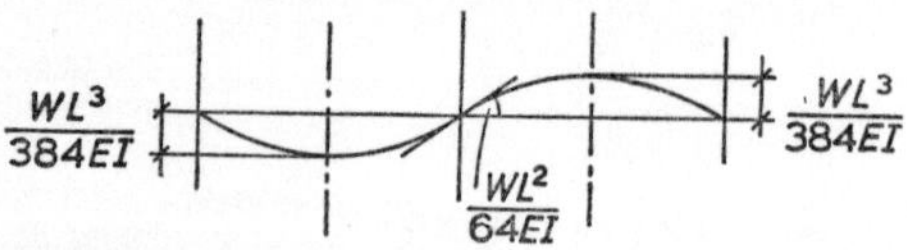

4.4

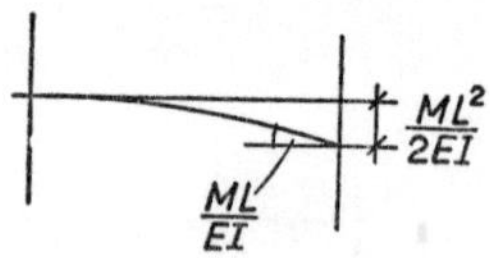

4.5

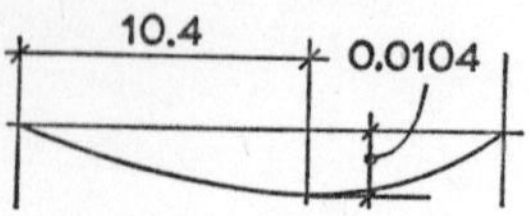

4.6

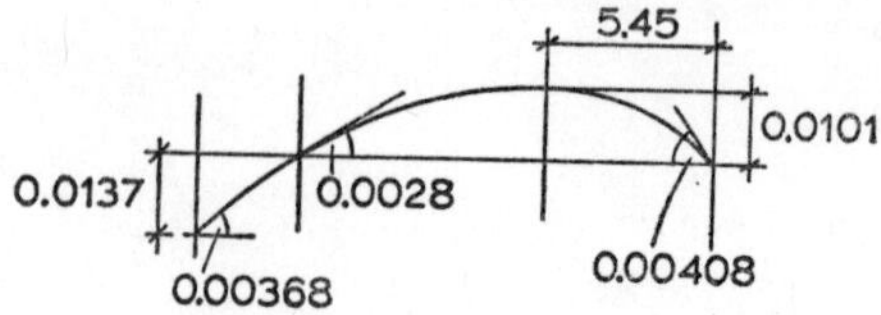

4.7

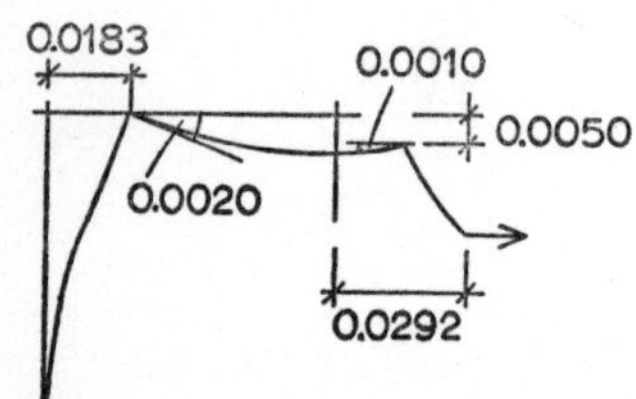

4.8

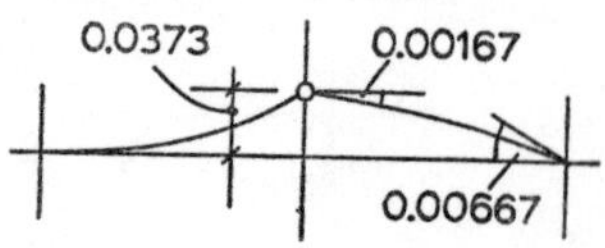

4.10

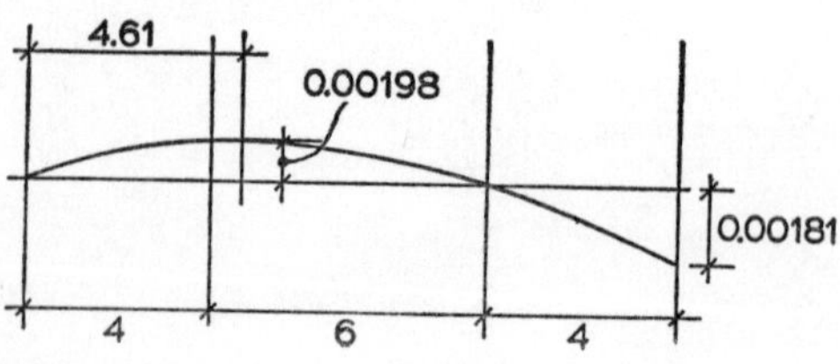

4.11 0.003 79

4.12 0.004 41, 0.034 4

4.13

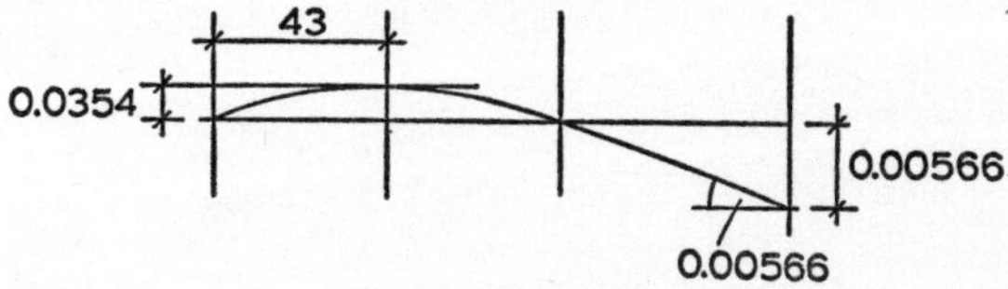

4.14

$$\begin{bmatrix} \frac{12}{L^2} & \frac{6}{L} & -\frac{12}{L^2} & \frac{6}{L} \\ \frac{6}{L} & 4 & -\frac{6}{L} & 2 \\ -\frac{12}{L^2} & -\frac{6}{L} & \frac{12}{L^2} & -\frac{6}{L} \\ \frac{6}{L} & 2 & -\frac{6}{L} & 4 \end{bmatrix} \frac{EI}{L}.$$

INDEX